The Playful Brain

The Playful Brain

The Surprising Science of How Puzzles Improve Your Mind

RICHARD RESTAK, M.D.,

with puzzles by SCOTT KIM

RIVERHEAD BOOKS

New York

RIVERHEAD BOOKS
Published by the Penguin Group
Penguin Group (USA) Inc.
375 Hudson Street, New York, New York 10014, USA
Penguin Group (Canada), 90 Eglinton Avenue East, Suite 700, Toronto, Ontario M4P 2Y3, Canada
(a division of Pearson Penguin Canada Inc.)
Penguin Books Ltd., 80 Strand, London WC2R 0RL, England
Penguin Group Ireland, 25 St. Stephen's Green, Dublin 2, Ireland (a division of Penguin Books Ltd.)
Penguin Group (Australia), 250 Camberwell Road, Camberwell, Victoria 3124, Australia
(a division of Pearson Australia Group Pty. Ltd.)
Penguin Books India Pvt. Ltd., 11 Community Centre, Panchsheel Park, New Delhi—110 017, India
Penguin Group (NZ), 67 Apollo Drive, Rosedale, Auckland 0632, New Zealand
(a division of Pearson New Zealand Ltd.)
Penguin Books (South Africa) (Pty.) Ltd., 24 Sturdee Avenue, Rosebank, Johannesburg 2196,
South Africa

Penguin Books Ltd., Registered Offices: 80 Strand, London WC2R 0RL, England

While the authors have made every effort to provide accurate telephone numbers and Internet addresses at
the time of publication, neither the authors nor the publisher assumes any responsibility for errors, or for
changes that occur after publication. Further, the publisher does not have any control over and does not
assume any responsibility for author or third-party websites or their content.

First Riverhead hardcover edition: December 2010
First Riverhead trade paperback edition: December 2011
Riverhead trade paperback ISBN: 978-1-59448-545-9

The Library of Congress has catalogued the Riverhead hardcover edition as follows:

Restak, Richard M., date.
   The playful brain : the surprising science of how puzzles improve your mind /
Richard Restak and Scott Kim.
   p.   cm.
   Includes bibliographical references and index.
   ISBN 978-1-59448-777-4
1. Cognitive neuroscience.   2. Puzzles.   3. Thought and thinking—Problems, exercises, etc.
I. Kim, Scott.   II. Title.
QP360.5.R47   2010   2010024140
612.8'2—dc22

PRINTED IN THE UNITED STATES OF AMERICA

10   9   8   7   6   5   4   3   2   1

To Toby and Mr. B

CONTENTS

Cognition

INTRODUCTION

We hear a lot these days about enhancing brain performance with mental exercises like crossword puzzles and the currently popular Sudoku. My own interest in puzzles stems from my ongoing effort to develop new and innovative approaches to brain-performance enhancement. I've aimed at answering this question: What activities can my readers engage in that will enhance not just the whole brain but *distinct brain areas and processes*? In *Mozart's Brain and the Fighter Pilot,* I came up with a program based on twenty-eight suggestions. None of these involved puzzles. In my second brain-enhancement book, *Think Smart: A Neuroscientist's Prescription for Improving Your Brain's Performance,* I asked some of the world's most prestigious brain scientists to tell me the specific activities they personally engage in to make their brains work better. During interviews and conversations with those neuroscientists, I was reminded of the importance of puzzles as brain enhancers and, as a result, included a small number of puzzles in *Think Smart.* At that point I began to look a bit deeper.

Over the years I had always been fascinated with puzzles. They are a fun way to stimulate the brain. But can they actually improve different brain functions, such as memory, reasoning, and three-dimensional visualization, among others? Over the course of writing nineteen books on the brain, I encountered a lot of formal and informal research suggesting the answer to that question is yes. Gradually I became convinced that puzzles can help enhance specific brain functions and, as studies suggest, actually help ward off mental deterioration.

What I needed, I reasoned, was a brain-puzzle book that included puzzles of every sort accompanied by an explanation of the brain benefits that can be expected from the different ones. As far as I could determine, no such book was available. Sure, any number of brainteaser books existed that contained puzzles purported to challenge the brain. But none of them discussed what happened in the brain while solving the puzzles or provided any explanation of how different puzzles challenged specific brain areas. Nor did they suggest what kind of puzzle to select in order to improve specific brain functions. Although I saw the need for such a book and was enthusiastic about writing it, one huge impediment stood in my way: my strictly amateur status as a puzzle solver (forget altogether about my not being a puzzle designer). Since I could go only so far as an amateur puzzle aficionado, I needed to be in touch with a professional: a puzzle master.

Anyone with an interest in puzzles soon comes into contact with my favorite puzzle master, Scott Kim. For years I've been fascinated and intrigued with Scott's puzzle-creating genius. One afternoon, while working on a series of his puzzles I came up with the idea of an innovative performance-enhancement book based on puzzles. The book would combine my knowledge about the brain with Scott's puzzle-creating talents in order to help our readers increase their brain power via entertaining and instructive puzzle challenges. Scott, as I later learned, had been thinking along similar lines while writing his regular puzzle column for *Discover* magazine and maintaining his website, www.scottkim.com.

In 2006 my curiosity about the brain–puzzle connection led me to contact Scott. To my delight I discovered that he was familiar with my work as well as the research findings and published writings of many other neuroscientists. Most important, he was well aware that puzzles can be used as brain enhancers. After speaking to each other for several hours while attending a neuroscience meeting, we both came away convinced that by working in collaboration we could create a unique pathway to enhancing brain function through puzzles. That meeting was the genesis of this book.

When we finally got together to plan the book, we talked about how we could combine information about the brain with the fun and challenge of solving puzzles. Here's the plan we came up with: First, I would identify and write about those brain functions that inevitably undergo decline unless deliberate efforts are taken to enhance them. Included here are concentration, memory, fine motor skills, visual observation, logic, numbers, vocabulary, visual-spatial thinking, imagination, and creativity. Then Scott would develop puzzles aimed at engaging and challenging each of these mental functions. Like two jazz musicians working out a contrapuntal melody, we would take up each mental function. I would write about what we know of the relevant brain processing, and Scott would design puzzles aimed at engaging and challenging the brain areas responsible for that function. Obviously we couldn't cover everything about the brain, so we concentrated on areas that lent themselves most easily to puzzles. The chapters would be self-contained and, because they featured puzzles, fun to read in any sequence according to the interests of the reader—sort of like leisurely making a selection from a box of delicious candy.

Following are three examples, drawn from Scott's vast puzzle output, of the kinds of things we'll be covering in this book.

LETTER SWAP *(Word Thinking)*

ANSWER ON PAGE 11

▶ *This puzzle exercises your ability to recall words based on meaning and spelling.*

Look at the string of boxes in the diagram and place letters in the boxes to spell two words differing by only one letter, like CHANGE and CHANCE. The final letter H in one word will become a P in the other word.

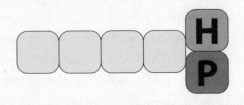

Readers who deal primarily with words rather than images or abstract concepts will more easily solve this puzzle, which challenges the left, language-mediating hemisphere of the brain. Since Scott has identified this as a hard puzzle, perhaps I should give you a hint: "Hold hands or fight." If you still haven't gotten it, here's another hint: The first letter is C. The solution is on page 11.

In chapter 9, on words and language, I'll use puzzles like this, combined with some of the newest positron emission tomography (PET) and functional magnetic resonance imaging (fMRI) findings, to explain how the brain is organized for language, the importance of lifetime vocabulary building, and how to use language to think creatively. Each of the points will be illustrated by puzzles and exercises such as the one you just solved.

As an immediate follow-up to that word puzzle, let's try a puzzle that challenges a completely different part of the brain. Tests of spatial thinking are especially brain-performance-enhancing for those of us who aren't architects or designers because they force us to use parts of our brain that we don't usually call upon. Here is one of Scott's puzzles that can be solved only visually-spatially:

PAPERWORK *(Spatial Thinking)*

ANSWER ON PAGE 11

▶ *This puzzle exercises your ability to arrange items in space according to rules.*

Lay six sheets of paper flat on a table so that each sheet overlaps above exactly one other sheet, and overlaps below exactly one other sheet. The sheets may not be folded, cut, or bent.

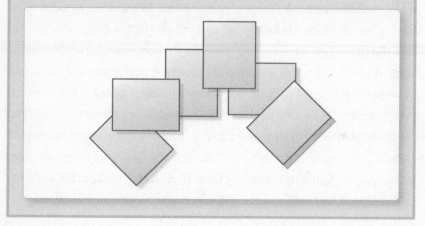

No amount of verbal reasoning can lead to the solution to that puzzle. The correct answer can come only from mentally moving the sheets in one's mind. Hint: Picture the papers joining together to form a simple geometric form.

If you're like me and your work primarily involves writing or reading rather than visual-spatial thinking, you probably will have a hard time with that puzzle. However, the brain benefit doesn't result from getting the correct answer (shown on page 11) but from the activation of the spatial-processing areas of your brain (the parietal lobes) that spring into action whenever you mentally envision shapes or locations. With practice in stimulating the parietal lobes via puzzles and exercises, you will be equipped to temporarily suspend your customary overreliance on words and language when you're faced with a puzzle or real-life challenge that calls for spatial thinking.

To illustrate this distinction between verbal and visual brain processing, imagine that you've become lost while driving; you have neither map nor GPS and you're in a hurry. Frustrated, you pull into a gas station and ask the attendant for directions. He wants to be helpful but he's busy. Would you prefer him to give you a verbal description of how to get to your destination, or would you prefer that he highlight the route on a map? People who work with words prefer verbal directions ("Drive two miles, turn left, drive one mile, and then make two successive left turns") over being handed a map and then having to mentally translate the directions on the map into spatial coordinates when back on the road. In chapter 4, on visual thinking, some of Scott's puzzles will help "language types" boost the power of their parietal lobes. Work on a sufficient number of these puzzles and you'll eventually be able to think like an architect or interior designer.

Probably the most important brain faculty worthy of enhancement is creativity: our ability to think "outside the box." On the next page is one puzzle that can't possibly be solved unless you overcome your initial preconceptions about a piece of paper.

MORE PAPERWORK *(Spatial Thinking)*

ANSWER ON PAGE 11

▶ *This puzzle challenges you to think creatively about creating a strong spatial structure.*

Balance a full wineglass at least four inches above the surface of a table using nothing but an ordinary sheet of stationery for support. Obviously, something has to be done with the paper, but what?

The solving of such puzzles requires the active participation of the right hemisphere of the brain, especially the frontal and anterior temporal lobes. These brain areas activate and coordinate a widespread system of brain circuits that aren't always working together. This coordination provides the necessary conditions for creativity first articulated in 1931 by famed experimental psychologist Charles Spearman: bringing together two or more ideas that previously had been unconnected.

In the puzzle above, the solution springs immediately to mind as soon as you combine the concepts of paper and columns. How does one transform a piece of flat paper into a column? Aha . . . of course! If you haven't solved it, read the solution at the conclusion of this section; otherwise, the next paragraph won't make much sense.

Puzzles requiring creativity for their solution are difficult to solve because the neural circuits that represent the various elements of the puzzle (a piece of paper; some means of using the paper to suspend the wineglass four inches in the air) have only weak connections with one another: we don't ordinarily use paper as a means of supporting glasses filled with wine. Therefore, our mental reliance on the "usual" way of thinking and doing things makes it unlikely that we will easily come up with the proper sequence required for the puzzle's solution. But if we can allow for spontaneous, even unbidden associations to arise while working on the puzzle, we can solve it.

Thanks to recent fMRI findings, we now know what's happening in the brain during that "Aha!" response we experience the moment we've solved an especially challenging puzzle. About 300 milliseconds before we become aware of the solution to the puzzle, a burst of activity occurs in the right hemisphere. This burst is interpreted by the authors of a pivotal fMRI study ("Neural Activity When People Solve Verbal Problems with Insight") as "making connections across distantly related information . . . that allow them to see connections that previously eluded them."

In the following chapters, Scott and I will take up the other key mental processes that can be enhanced via puzzles. We're aiming at a book that you will return to again and again. And as Scott mentions, we have set up a special website—www.theplayfulbrain.com—where new puzzles can be downloaded to further sharpen the brain processes we concentrate on in this book. In addition, we provide other sources where you can find similar puzzles.

Our overall goal is to help our readers make maximum use of each of their brain functions and thereby achieve what psychologists refer to as cognitive complexity: richer, increasingly nuanced perceptions, longer-lasting memories, and more accurate responses to people, situations, and events.

SCOTT While Richard came to this book as a neuroscientist seeking to work with a puzzle designer, I came from the other direction: as a puzzle designer seeking to work with a neuroscientist.

For years I have created puzzles for magazines, electronic games, and toys. I've created many different types of puzzles: number puzzles, visual puzzles, musical puzzles, educational puzzles, and yes, even the occasional crossword puzzle.

Puzzles exercise your mind the same way sports exercise your body. I believe that mental exercise should be considered a basic part of healthy living, along with physical exercise and good nutrition.

In recent years, the idea of playing games to exercise your brain has exploded in popularity.

First, the number-logic game Sudoku became a worldwide hit. Even people who hate mathematics were hooked by its seductive invitation to flex their logical-thinking muscles. People played it with a conscious belief that Sudoku would help them keep their minds sharp. Some doctors even prescribed Sudoku to their older patients.

At the same time, the electronic game Brain Age for the Nintendo DS exploded in Japan, rapidly rising to the top of the best-seller charts. Soon after, it conquered America and Europe. This canny collection of minigames put brain fitness front and center, ranking your mental performance after each session. Thanks to Brain Age, a fifty-year-old man can achieve the mental astuteness of a twenty-year-old. Who doesn't want a *young* healthy brain?

Finally, articles have flooded popular media claiming that playing games can help keep the mind sharp and perhaps even stave off Alzheimer's and other degenerative brain diseases. Recent advances in brain imaging show that even small amounts of mental exercise can stimulate neurons in certain parts of the brain to grow new connections. Brain degeneration is no longer an inevitable fact of aging. There is hope.

All this interest in games for health got me interested in creating puzzles for mental exercise. But exactly how do puzzles strengthen your brain? Which types of puzzles were best for strengthening which mental functions? Where was the scientific research? I realized I needed to talk with a brain scientist.

Fortunately, Richard found me first. We met at a neuroscience conference in San Diego, and I found him to be a most congenial collaborator. Our visions of what we needed to do meshed perfectly. As Richard explained in his part of this introduction, our goal was to write a book that included both puzzles to play and the science to back them up.

While I worked with Richard, I learned something surprising about the state of neuroscience. I've always been well aware that the brain is a complex and subtle organ and that, although we have made much progress, we are just beginning to fathom how the brain works. What I didn't know is that most brain research in the past has focused on sick brains and how to fix them, and not on healthy brains and how to maintain them. There is surprisingly little research on brain exercise for healthy people.

Undoubtedly a big reason for this imbalance has been that in the past the only way to figure out which areas of the brain did what was to observe patients who had lost parts of their brains. Only recently have modern imaging techniques allowed us to observe moment-to-moment brain changes in a healthy person.

Now a few words about the puzzles here . . . Throughout this book there are many puzzles for you to try. Each group of puzzles targets a particular function of the brain: memory, visual thinking, spatial reasoning, logical reasoning, and so on. Some of the puzzles were inspired by scientific research; others I based on activities that teachers use to introduce different styles of thinking.

I encourage you to try all the puzzles, even if some of them look hard or unfamiliar. No one is grading your work, so don't worry about getting all the answers right. Just give each puzzle a try and enjoy seeing how far you get. Trying a puzzle will reward you whether or not you finish it.

For instance, the puzzle about sounds in chapter 6, on listening (page 149), asks you to match sound effects with the objects that produce them. Whether or not you figure out the correct answers, you will find that focusing your attention on sounds and how they are produced will increase your awareness of other sounds in your environment.

Each group of puzzles includes several puzzles that start easy and get harder. If you have trouble getting started, feel free to jump straight to the answers. I've written answers that explain how to solve the puzzles, so reading the answer to one puzzle will help you solve the next puzzle.

Finally, have fun. The more you enjoy using your brain, the more you'll get out of it. Enjoy sampling the morsels of "brain candy" in this book. We hope it whets your appetite for more.

ANSWER TO LETTER SWAP

ANSWER TO PAPERWORK

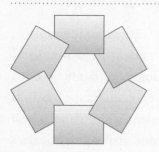

ANSWER TO MORE PAPERWORK

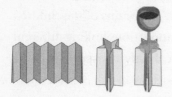

The Art and Science of Brain Enhancement

The most profound insight into the brain achieved in the late twentieth century was the discovery that, over a person's life span, his brain never ceases to undergo structural and operational changes based on life experiences. The richer those experiences, the greater the brain's development. Brain scientists use the term *plasticity* to describe this process, and they first observed it in experimental animals—rats, to be specific.

If a rat is raised in an "enriched" environment—for example, a cage full of toys and gymlike equipment, along with other rats to play with—its brain will develop a greater number of nerve cell connections (synapses) and increased nerve fiber (dendritic) complexity, especially in the hippocampus, where memory is initially encoded. As a result of possessing more complex brain circuitry, environmentally enriched rats perform better on tests measuring learning and spatial memory. In one test of spatial memory, the Morris water maze, the rats are placed in a small pool of murky water in which they must swim until they find a submerged escape platform. In this test, environmentally enriched rats consistently outperform rats raised under standard, nonstimulating environments.

Enhanced brain development and performance occur in monkeys, cats, and a host of other animals raised in enriched environments. Does the same rule hold for humans? Almost certainly it does, neuroscientists are convinced, although they cannot, for obvious ethical reasons, design a comparative experiment with infants to test that conclusion. But neuroscientists have found striking deficiencies in intellectual and emotional development among infants raised in institutions compared with infants of the same age who are transferred from the institutions into adoptive families. The brains of the children who remain in the institutions have fewer connections linking different parts of the cortex as well as reduced chemical activity, especially in the frontal and temporal areas—two sites important in IQ, memory, and other cognitive functions.

In adults, experiments have been carried out on individuals who have spent years perfecting specific skills. The reasoning underlying these experiments goes like this: thanks to its plasticity, the brain should show enhanced development in those areas used for the performance of a specific skill.

THE BRAINS OF LONDON CABDRIVERS

The most convincing study confirming this hypothesis was carried out in 2000 on London cabdrivers. In contrast to what occurs in most American cities, where cabbies are given little formal instruction and largely learn the geography of a city while on the job (often with the help of a GPS system), aspiring London cabdrivers spend up to three years learning the intricacies of London's oftentimes bewildering labyrinth of streets. After three years of intense study, they must pass a demanding exam that tests their ability to drive from one address to another via the shortest path.

Among London cabbies, the hippocampus—a sea-horse-shaped structure important in spatial learning and memory both in humans and in the rats exposed to the water maze mentioned earlier—is significantly larger than in people with less familiarity with the geography of the city. What's more, the size of the posterior part of the hippocampus varies according to the number of years the cabbies have been driving. In those with a decade or more of experience, the posterior hippocampus was found to be larger than in cabbies with only a few years of experience.

YOU CAN IMPROVE YOUR BRAINPOWER

This plasticity-enriching brain enhancement continues until the day we die. By learning more, doing more, and experiencing more, we form greater numbers of circuits within the brain and thereby increase its functional power. As a result, we become smarter, faster, and more astute.

L

Whenever we speak of ways to improve our brain's functioning we're usually referring to *cognition*, the process by which we attend, identify, and react to both the external world and our own thoughts. In order to get a firm grip on what is meant by cognition, consider this: only 25 percent of the brain is devoted to encoding all of the information coming to us through all of our senses. The remaining 75 percent integrates all that incoming information and transforms it into the unified world of our experience.

Look for a moment at the diagrams below, which illustrate that 75 percent/25 percent breakdown. As each of the brain's major lobes carries out specific functions (sight, hearing, etc.), its association areas project outward to commingle with the association fibers streaming from other lobes, much like a network of rivers emptying into the ocean. Association circuits within each lobe convey specific information. The parietal association areas direct our attention to the location of what's going on around us; the temporal association areas help us to interpret what we observe; the frontal association areas help us to appropriately respond. We know these facts about the brain thanks to the study of unfortunate individuals who have suffered damage to their parietal, temporal, or frontal areas.

LEFT HEMISPHERE-Surface

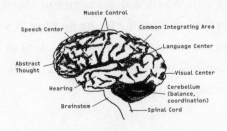

RIGHT HEMISPHERE-Interior

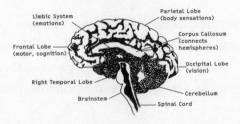

People with injuries to their parietal lobes often lose the ability to pay attention to objects—including their own bodies—in space. Ask them to draw a clock, and they will leave out the half of the clock (either the right or left side) ordinarily processed by the parietal lobe on the opposite side of the brain. (No one has satisfactorily explained the brain's odd anatomical arrangement whereby information to and from one side of the body is processed on the opposite side of the brain.) Damage to the temporal association areas results in problems recognizing, identifying, and naming. Show a person with temporal-lobe damage a familiar object—on occasion, even his own picture— and he will fail to recognize it, a defect referred to by neurologists as an agnosia, from the Greek for "not knowing." Finally, frontal-lobe-associated damage leads to the most devastating impairments of all: loss of the power to balance past and present experiences with future expectancies. As a result, a person with frontal-lobe damage has difficulty planning and organizing his life and is often described by others who used to know him well as "a changed person."

In the normal brain, the association fibers from different parts of the brain mutually influence one another, and the contributions of all of the association areas add up to more than the sum of their parts. Think of the association areas as a vast interconnected network that has the potential to link every part of the brain. When it comes to simple sensation, this linkage is especially easy to observe. We hear our name called and turn in the direction of the voice we hear; we feel something crawling on our forearm, look down at the fly responsible for this sensation, then swat it away.

Similar integration occurs with thoughts, concepts, and ideas. Thanks to the association areas, everything new that we learn has the potential to be linked with all of the things that we already know. As a corollary, the more we know, the more highly evolved our association areas become, thanks to increases in the brain cell connections within the association areas. But how does this come about at the level of individual nerve cells? The Canadian psychologist Donald O. Hebb

B

suggested the most commonly accepted explanation in his now classic 1949 book *The Organization of Behavior*.

Learning, according to Hebb, entails strengthening certain connections among some cells at the expense of others. When we learn new information or perfect specific skills, we strengthen circuits composed of interlinking neurons. "Cells that wire together fire together" and "Use it or lose it" are oft-quoted neuroscientific mantras that sum up the implications of Hebbian learning. And because each person's life experience is unique, no two brains—not even the brains of identical twins—are exactly alike. But a mystery lurks here, one that is informally referred to by philosophers and neuroscientists as the "binding problem": we experience people and objects as unities even though we use different sensory channels to perceive them.

For instance, when we go to a baseball game, we don't experience it as separate inputs coming from our eyes (the players), ears (the cheering crowd), and bodily senses (the uncomfortable seats). Thanks to the association areas, everything is bound together in our brain into a single experience—including our emotional response (disappointment) when our favorite batter strikes out. The binding of our sensations persists into the future too: the aroma of hot dogs reminds us years later of those magic afternoons we spent at the ballpark.

But how is the firing of neurons and neuronal circuits responsible for so many sensory experiences—size, shape, texture, color, smell, taste—synthesized into a whole, i.e., the dish of ice cream that I'm now eating? Different answers to that question have been suggested. Perhaps the brain contains distinct cells or small circuits of cells capable of responding to ice cream? But if ice cream cells or circuits exist, what happens in the brain to help us differentiate chocolate, strawberry, and vanilla ice cream? A better explanation involves synchronization of the firing times of a critical mass of different brain cells widely scattered throughout the brain that, when integrated, enable us to distinguish one ice cream flavor from another. In this model, temporal as well as spatial factors bind the different components of my ice-cream-eating experience.

Thanks to the vast interconnected network of the association areas (each nerve cell has input from up to 100,000 other nerve cells), the human brain has evolved powerful functions that are either not present in other species or are present in less powerful forms. Included here are language, memory, perceptual and motor-skill learning, reasoning, visual thinking, logic, creativity, and mathematics, among others. The exciting news about each of these functions is that they can be improved by puzzles.

Each of the following chapters will explore a different brain function. After describing the workings of the relevant brain areas, we'll suggest ways to strengthen that function by working at puzzles. The first and arguably most important function is working memory. But before tackling working memory and the other brain functions, we provide some general principles governing how we go about solving puzzles. That way, you can segue smoothly from learning about the brain to working on puzzles that will strengthen brainpower.

Learning to Solve Puzzles

The puzzles in this book are a fun way to exercise your mind and get your brain to work more efficiently. You will find a wide variety of puzzles that focus on many different areas of your brain. Some puzzles will exercise skills you already have, while others will open your mind to unfamiliar ways of thinking.

Solving puzzles is a skill you can learn. When you first start solving crossword puzzles, for instance, words come slowly and it is easy to get stuck. But the more you persist, the more familiar you become with the peculiar words that crop up often in crosswords and the more nimbly you hop around the grid, filling in letters. Similarly, you can learn techniques to improve your performance on jigsaw puzzles, logic puzzles, and Sudoku.

Although different types of puzzles require different problem-solving tactics, there are some basic strategies that will help you solve all types of puzzles. Learning these universal problem-solving strategies will not only help you solve puzzles faster, but also help you solve problems in your everyday life.

Here are my top ten tips for solving puzzles. I encourage you to try them out as you read this book, especially when you get stuck. Not only will you get more out of the puzzles, you'll have more fun as well.

1. READ THE INSTRUCTIONS TWICE.

What is the goal of the puzzle? What are you allowed to do and not allowed to do? Most puzzles in this book have straightforward instructions, but tricky puzzles—like the ones in chapter 14, "Creativity"—have deceptive instructions that lead you into making false assumptions.

Here is a classic riddle with deceptive wording:

AS I WAS GOING TO ST. IVES *(Word Thinking)*

▶ *This puzzle challenges you to read the problem carefully.*

As I was going to St. Ives,
I met a man with seven wives,
And every wife had seven sacks,
And every sack had seven cats,
And every cat had seven kits.
Kits, cats, sacks, wives,
How many were going to St. Ives?

Most people on hearing this riddle for the first time get caught up in counting how many wives, sacks, cats, and kits there were altogether. In a conventional puzzle, this would indeed be the core of the question. But in this puzzle, the multiplication problem is a big red herring. The key is to realize that "As I was going to St. Ives" implies that only the speaker was actually going to St. Ives. The answer is: Just one. (To be honest, the logic here is not airtight, since it is possible to meet someone who is going the same direction.)

Mystery novelists perform similar sleight of hand when they disguise the clues to solving a crime as apparently irrelevant details in the story. You will find other similarly misleading puzzles in chapter 13, "Illusions," and chapter 14, "Creativity." To solve these devious dilemmas, you'll need to read the instructions carefully.

2. ASSESS THE CHALLENGE.

After you read the instructions, size up the situation. Is the puzzle hard or easy? Interesting or boring? Familiar or strange? Intimidating or inviting? Be honest. Your emotions will tell you a lot about how you need to prepare yourself in order to solve the puzzle.

If a puzzle seems uninteresting, imagine you are on a television game show. You have already answered several questions correctly and feel happy and excited. Answering this question correctly will win you a big prize. Now what would you do?

If a puzzle seems too hard, imagine you are an actor playing the part of the world's best puzzle solver. How would that person approach the puzzle? Suppose you have a team of assistants to help you. Who would you call on?

Following are two puzzles. Don't solve them yet—we'll get to that in a moment. For now, just notice how you feel about them. Chances are the two puzzles will evoke different feelings.

A

PIGPENS *(Spatial Thinking)*

ANSWER ON PAGE 34

▶ *This puzzle challenges you to arrange objects in space according to rules.*

Draw three straight lines to separate the seven pigs into seven separate pens.

THREE DICE *(Number Thinking)*

ANSWER ON PAGE 34

▶ *This puzzle exercises your ability to think about numbers and arithmetic.*

Jack just rolled three dice. When you multiply the three top numbers together, you get a result that is twice as large as when you add the three numbers together. What three numbers did Jack roll? There is more than one correct answer.

1

3. TRY SOMETHING.

Sometimes just getting started can be the hardest step in solving a puzzle. If you don't know what to do, don't worry about solving the whole thing. Just get started. Try something. Anything. Even a wild guess is fine. Then check your guess to see whether it helps you solve the problem. Often your guess will lead to a dead end, but that's okay. Understanding why a guess doesn't work can help you figure out where to focus your efforts next. Let's try making a guess on the Pigpens and Three Dice puzzles.

PIGPENS

Suppose you draw a line as shown below. One pig is now in its own pen, but after some experimenting, you find that you are not able to divide the remaining six pigs into six pens using just two more lines. That tells you that the first line should divide the pigs into two groups that are of more equal size.

THREE DICE

You have no idea of how to get started, so you guess the numbers 4, 5, and 6. Now check. The sum of the three numbers is 15. When you multiply the three numbers together you get 120, which is much bigger than what it should be (twice 15). That tells you that your next guess should probably involve smaller numbers.

4. PERSIST.

Solving a puzzle takes effort. So don't give up. If you get discouraged, or worry that you aren't doing well, remember that no puzzle is easy until you have solved it. Keep trying and you'll make progress. Expert puzzle solvers don't take failure personally. Instead, they simply say "That didn't work" and try again, drawing on their arsenal of techniques for recovering quickly from mistakes.

Some puzzles take more persistence than others. Puzzles such as N-back in chapter 1, "Working Memory," and Triangle vs. Square in chapter 7, "Motor Skill Learning," ask you to perform tasks that take time and practice to master. Expect to put some time into these.

Can you persist and solve the Pigpens puzzle? To get you started, here is the first line. Keep trying until you get it. The answer is on page 34.

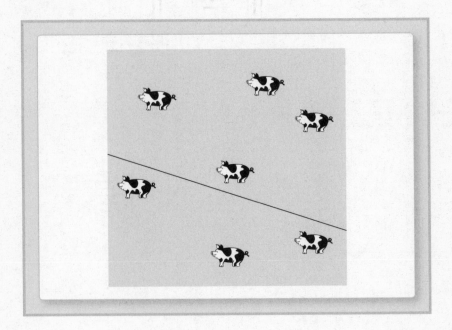

To test your ability to persist, here's a big maze. Start at the top and end at the bottom. Can you persist and make it through?

MAZE *(Spatial Thinking)*

ANSWER ON PAGE 35

▶ *This puzzle challenges your persistence and ability to follow paths in space.*

START

FINISH

B

5. BE SYSTEMATIC.

Some puzzles can be solved through intuition and flashes of insight. Other puzzles, like those in chapter 10, "Logic," require a more disciplined approach. When solving a puzzle that requires considering many possibilities, come up with an orderly way to step through the problem that marches steadily toward a solution.

For instance, how many four-letter words end in the letters UST? If you guess letters at random, you will not be sure you got all the answers. A better approach is to systematically step through the letters of the alphabet (AUST, BUST . . .), which lets you be certain you got all the answers.

Here's another puzzle that benefits from a systematic approach.

SQUARE COUNT (*Visual Thinking*)

ANSWER ON PAGES 35–36

▶ *This puzzle exercises your ability to find patterns within a larger figure.*

How many squares of any size are in the figure below?

For many puzzles, it helps to draw a chart of your progress. Here is a chart that helps you solve the Three Dice puzzle (page 21) systematically. Remember that the directions don't say you can't repeat numbers.

A	B	C	A × B × C	A + B + C
1	2	3	6	6
1	2	4	8	7
1	2	5	10	8
1	2	6	12	9
2	2	2	8	6
2	2	3	12	7
2	2	4	16	8 (solution!)

Now, how could you systematically solve the Square Count puzzle?

6. BE EFFICIENT.

Figure out what part of the puzzle is most vulnerable to attack. Where can you make the most progress with the least effort? For jigsaw puzzles, the best way to start is by assembling edges: they're easier to find, they're easier to assemble, and they let you make progress quickly.

Whenever you look at a puzzle, look for what is easiest to figure out first.

Where should you start in the puzzle that follows?

CAN YOU DIGIT? *(Mathematical Thinking)*

ANSWER ON PAGE 37

▶ *This puzzle exercises your ability to think about numbers and arithmetic.*

Put the digits 0 to 9 in the ten square boxes to make a correct sum. I've already placed three of the digits.

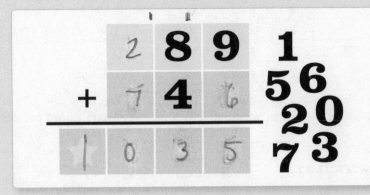

The most efficient place to start here is with the starred square, which can only be a 1, since the largest possible sum in the hundreds place is 19, or 9 + 9 + 1 carried from the tens place. That in turn implies that the digits in the hundreds place must add up to at least 10.

Next, consider the places where the 0 could go. Let's start with the ones column. Could the 0 go just below the 9? No, because then we would have 9 + 0 = 9 in the ones place, and we know that no digit appears more than once. Could the 0 go two squares below the 9? No, because then we would have 9 + 1 = 0 in the ones place, which would mean the digit 1 appears more than once.

Now let's move to the tens column. Could 0 go below the 8 and 4? No, the only numbers that can appear below 8 and 4 are 2 or 3, because 8 + 4 will either be 12 or 13 (if there is a carry from the ones place).

If 0 can't go in the ones place or the tens place, then it must go in the hundreds. From here, the puzzle unravels quickly.

7. SIMPLIFY.

If a puzzle is too hard or complex, try solving a simpler problem.

One way to simplify a puzzle is to break it into several smaller puzzles. For instance, you can break a crossword puzzle into several smaller puzzles linked only by a few words.

Another way to simplify a puzzle is to consider a simpler version of the same question. Consider the following counting puzzle.

SEATING ARRANGEMENTS *(Mathematical Thinking)*

ANSWER ON PAGES 30–31

▶ *This puzzle exercises your ability to systematically enumerate combinations.*

Four people (we'll call them A, B, C, and D) sit down to eat every night in four different chairs. They want to sit down in a different combination of chairs each night. How many nights can they sit down without repeating a combination?

The number of combinations is too big to imagine in your head. So ask a simpler question: How many nights can three people sit in a different combination of three chairs each night? It's not hard to list all six combinations—

ABC
ACB
BAC
BCA
CAB
CBA

—and this fact helps you solve the larger puzzle. Consider the starred chair. If person A sits in the starred chair, there are six ways to seat the remaining three people in the remaining three chairs. Similarly, if B sits in the starred chair, there are six ways to seat the remaining three people. The 4 possible people in the starred chair times 6 ways to seat the remaining three people = 24 combinations.

The Pentominos puzzle in chapter 5, "Spatial Thinking," asks you to find how many shapes can be made out of five squares. This puzzle is quite hard. A simpler version is to count how many shapes can be made with just four squares. An even simpler version is to count the shapes for just three squares. Once you solve these simpler puzzles, you will be better prepared to solve the full puzzle.

8. DRAW A DIAGRAM.

If you have trouble keeping track of all the elements of a puzzle, try representing it in a different way. Draw a diagram. Say it out loud. Write it in words. Gesture it with your hands. Often you will find that changing the way a puzzle is represented will make it easier to solve.

For instance, how many different capital printed letters are made entirely of straight lines? If you are good at visual thinking, you can probably envision the letters and count the answer in your mind's eye. If visual thinking is not your strong suit, draw the letters and count.

Here's a diagram that will help you understand the solution to the puzzle Seating Arrangements:

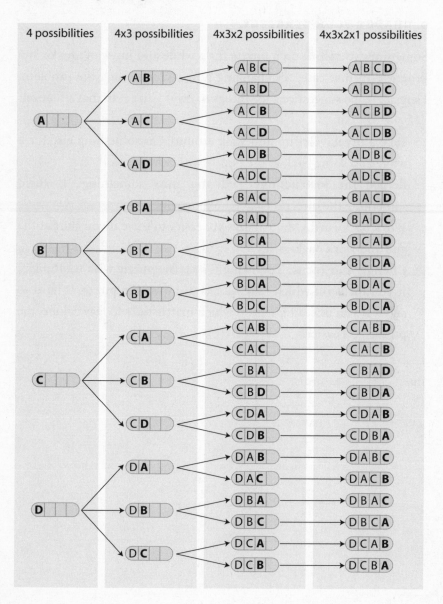

| 4 possibilities | 4x3 possibilities | 4x3x2 possibilities | 4x3x2x1 possibilities |

Sometimes it helps to translate a puzzle into other forms. The Clap Your Name game (in chapter 6, "Listening") is easier for some people if they speak the letters out loud. When I solve three-dimensional puzzles like Turn of the Century (in chapter 5, "Spatial Thinking"), I often gesture with my hands to get a more tactile sense of the shapes.

9. CHANGE YOUR STRATEGY.

Sometimes you work on a puzzle for a while and just can't make any progress. In that case, try changing how you approach the problem. Here are some suggestions for things to try if you get in that situation:

- **Take a break**. Get up and walk around. Do something else for a while and come back to the puzzle with a fresh mind.
- **Reread the instructions**. Did you miss something? Is there something the directions aren't telling you?
- **Work backward**. Many mazes are easier to solve if you start at the end and work backward.
- **Change your focus**. Following is a classic puzzle with toothpicks. If you focus on which matchsticks to move, the puzzle is hard to solve. If you instead focus on which matchsticks to leave alone, the puzzle gets easier.

ANSWER ON PAGE 37

▶ *This puzzle exercises your ability to think creatively about arranging things in space.*

Move just two matchsticks to rebuild the goblet so the olive is outside the goblet.

10. GET HELP.

If all else fails, read the answer. When I first started reading puzzle books, I often turned straight to the answer. That may sound like cheating, but I learned a lot by taking the time to understand why the answer was correct and imagining how I might have gotten the answer.

I've written answers in this book so reading the answer to one puzzle will help you solve the next. I encourage you to read the answers, especially if you find a puzzle too frustrating or if you want to compare your solution with mine.

~

Solving a puzzle is like climbing a mountain: when you first look up at a mountain, it may seem impossible to reach the top. Some mountains

A

are gentle slopes, while others are steep cliffs that require equipment and preparation. When you are at the bottom, it is hard to see the best path to the top. And when you reach the top, you experience the thrill of knowing you lifted yourself up using your own energy.

With this as background, the next chapter explores some of the facts that neuroscientists (brain scientists) have recently learned that provide insight into what takes place in our brain when we solve puzzles. Learning about the brain confers a dual advantage: we can use what we learn about the brain to help devise strategies for successfully solving puzzles; at the same time, by working at the puzzles, we enhance our brain's efficiency and power. In chapter 1, we'll start by taking up what is probably the most important of our brain's faculties: working memory.

ANSWER TO PIGPENS

Here is how to divide the seven pigs into seven separate pens with three straight lines.

ANSWER TO THREE DICE

There are two correct answers:

1 4 5 (1 x 4 x 5 = 20, which is twice 1 + 4 + 5 = 10)
2 2 4 (2 x 2 x 4 = 16, which is twice 2 + 2 + 4 = 8)

The gray line shows the solution path.

ANSWER TO SQUARE COUNT

It's easy to see that there are sixteen small squares here. But there are also larger squares of other sizes. Again, if you count them randomly, you will not be sure you got all the answers. A better approach is to count all the one-by-one squares, then all the two-by-two squares, then three-by-three, then four-by-four. And for each size, count the squares by pointing to the upper left corner of each square, scanning them in order from left to right, top to bottom. If you count the squares this way, you will see that there are 16 one-by-one squares, 9 two-by-two squares, 4 three-by-three squares, and 1 four-by-four square, for a total of $16 + 9 + 4 + 1 = 30$ squares.

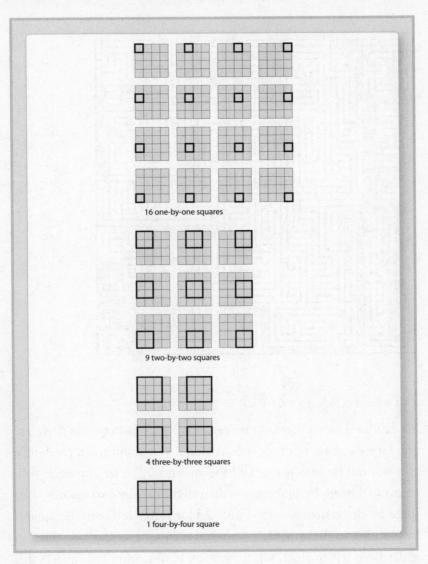

16 one-by-one squares

9 two-by-two squares

4 three-by-three squares

1 four-by-four square

ANSWER TO CAN YOU DIGIT?

$$289 + 746 = 1035$$

$$789 + 246 = 1035$$

ANSWER TO MATCHSTICK GOBLET

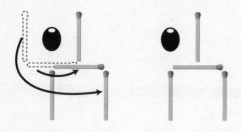

Y

Memory

1

WORKING MEMORY:
BRAIN JUGGLING

THINK BACK to the last time you lost your car key somewhere in your house. As soon as you realized it had gone missing, you began looking for it in a more or less systematic manner, starting with the location where you remembered last seeing it. If you didn't find it there, you widened your search, taking care not to revisit places you had already searched. Throughout your search, you remained focused on the goal of finding the key and ignored or short-circuited distractions ("Hi, thanks for calling. Can I call you right back? I'm busy with something that simply can't wait").

While carrying out the key search, you were exercising *working memory*, a feature that distinguishes the human brain from the brains of all other animals. As you combed through your house or apartment looking for the key, your brain was juggling all kinds of related information: what the key looked like; where you customarily put it; why it wasn't in its usual place; whether you had an extra key, and if so, where that might be; the consequences of arriving late for work if you invested too much time searching for the key; and finally, whether it might be better to call off the search, telephone a coworker, and ask him to pick you up on his way to the office.

Working memory is often metaphorically compared to juggling. A good juggler simultaneously keeps a varying number of balls in the air. Working memory is like that. It involves a relatively small number of items (averaging three or four for visual working memory) that are simultaneously kept track of. Just as some people can juggle more balls than other people can, the number of items in working memory varies from one person to the next.

Visual working memory is flexible and can be improved by effort. Simply spending more time looking at something or directing one's attention to it increases the chances of remembering it later. In one experiment to test this, subjects looked at random items flashed on a screen and later tried to recall the position and orientation of a specific item. The precision of their memory could be enhanced both by increasing the amount of time they spent looking at the item and by directing more attention to it during their initial scan of the items: noticing is remembering. This holds true not just for visual working memory but for working memory in general. You would not have lost your key if you had paid greater attention to where you last placed it. If you had done that, the location of the key would stand out in your working memory from all other locations.

A specific area of the brain is principally involved in working memory: the prefrontal cortex, the most anterior (farthest to the front) portion of the frontal lobes. (The parietal lobes also play a part.) In humans, the frontal lobes are among the last structures to mature; they encompass almost a third of the entire cortex. A localized portion of the prefrontal cortex, the dorsolateral prefrontal cortex, is especially involved in working memory. When activated, the dorsolateral prefrontal cortex forms an internal representation (a mental snapshot of that key), helps you remain focused during the search for it, and enables you to recognize it when you finally find it.

Since children and adolescents possess undeveloped prefrontal cortices—they aren't fully matured until early adulthood—their working memory isn't very efficient. They experience problems organizing themselves, keeping their attention focused, and managing more than one or two things at a time. In old age, especially among people with degenerative brain disease of the frontal lobes, working memory suffers along similar lines.

Why Working Memory Matters

What makes working memory so important is the mounting evidence that an optimally functioning working memory is the most important component of enhanced intelligence. Students with low working-memory skills are prone to misunderstandings and mistakes on tests measuring reading comprehension. In general, the higher a student's scores on the verbal portion of the Scholastic Aptitude Test (SAT), the better his working memory. A similar relationship exists between working memory and IQ as measured by standard nonverbal tests.

The relationship between IQ and the ability to solve geometrical puzzles forms the basis for the Raven's Progressive Matrices tests. According to their developer, Dr. John C. Raven, the tests are free of cultural or educational bias: people taking the tests aren't limited by information that they have learned in the past. No special information is needed to complete the sequence of geometric patterns. Further, the difficulty of each test is directly related to the number of factors that must be kept in mind (e.g., in working memory) in order to solve it. The more factors involved, the greater the stress on one's working memory ability and the harder it is to come up with the correct selection.

As you work at the following puzzles, you will be engaging the same prefrontal areas that are engaged when you take IQ tests or any exercise in working memory.

B

COMPLETE THE MATRIX *(Logical Thinking)*

ANSWER ON PAGE 45

▶ *This puzzle exercises your ability to understand pattern rules. It also exercises memory.*

In each puzzle below, choose the figure that belongs in the blank. Notice how puzzles 2 and 3 put more and more strain on your short-term memory.

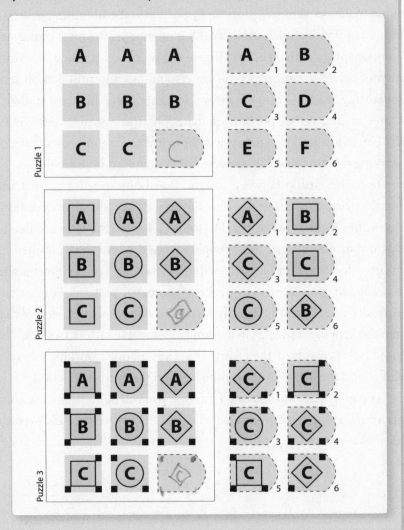

ANSWERS TO COMPLETE THE MATRIX

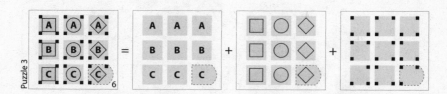

Puzzle 1 has a simple pattern: in every column the pattern is A, B, C.

Puzzle 2 adds a second pattern: in every row the pattern of shapes is square, circle, diamond. To solve this puzzle, you must remember two different patterns, which are shown separately above.

Puzzle 3 adds a third, more complex, pattern of black squares in the corners. To solve this puzzle, you must remember three different patterns, which are shown separately above.

A

Exercising Your Working Memory

COIN COUNTING *(Memory)*

▶ *This puzzle exercises your working memory as well as your mathematical ability.*

Gather a handful of pennies, nickels, dimes, and quarters and place them on a desk or table in front of you. You shouldn't count the number of coins ahead of time, but the desired number is anywhere between ten and fifteen of each, laid out in no particular order. Now pick them up one at a time and count the number of coins of each denomination. How did you do it?

Unless you have a highly developed working memory, you picked up one denomination at a time and totaled it before moving on to another denomination. It's unlikely you elected to pick up the coins at random while keeping a separate mental tally for each denomination and then totaling everything at the end. That's because it requires greater mental effort to keep track of all of the different denominations simultaneously. It's easier to

▶ ▶ ▶

▶ ▶ ▶

count the coins one denomination at a time because that process makes less of a demand on working memory. After counting the nickels, for instance, that total can be stored in working memory and your attention shifted to the next denomination.

In order to increase your working memory, count the pennies and nickels at random, i.e., don't alternate them. This will require you to keep a running total of each denomination in working memory while you're counting. Do it rapidly, and as you count each coin, discard it by placing it to the side. When you're finished, write down your totals and then check for accuracy by separately counting each of the denominations among the discarded coins. Not much practice should be required for you to manage two denominations.

Next, count three denominations in the same manner, and then, simultaneously, all four denominations. If you can manage four, you are achieving what psychologists consider the maximum number of items that can be stored in working memory. "Four items seems to be the limit. It's a fundamental characteristic of human working memory," according to Paul Verhaeghen, the psychologist at Syracuse University who carried out the experiments demonstrating the four-item limit. (If you want to read the details of how he determined this, check his fascinating paper "People Can Boost Their Working Memory Through Practice," published in the American Psychological Association's *Journal of Experimental Psychology: Learning, Memory and Cognition,* vol. 30, no. 6).

Here is another way of strengthening your working memory, known as the N-back Memory Game. You can play it using this book.

N-BACK GAME *(Memory)*

▶ *This puzzle strengthens your working memory.*

You will have to close this book before playing, so read these instructions thoroughly before you play the game.

1 Close this book. Turn to page 1. Notice the letter printed in the lower right corner of the page. Keep turning to the next right-hand page and notice the letter in the lower right corner.

2 Whenever you see the letter B (the trigger letter), try to remember the letter that appeared two pages before, and call it out. Then flip back two pages to see if your guess was correct.

3 Keep track of how many guesses you got right or wrong. With practice, you will see your performance improve.

4 Play the game again with a different trigger letter to prevent yourself from memorizing the location of the trigger letter.

5 For a harder challenge, try 3-back or 4-back: when you see the trigger letter, remember the letter that appeared three pages earlier or four pages earlier.

On pages 49 and 50 are other ways to play the N-back Game, using playing cards or a voice recorder.

N-BACK GAME WITH CARDS *(Memory)*

▶ *This puzzle strengthens your working memory.*

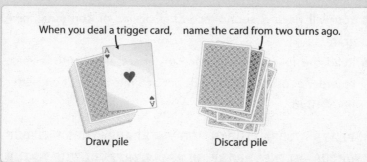

When you deal a trigger card,   name the card from two turns ago.

Draw pile   Discard pile

You can also play the N-back Game with a deck of playing cards.

① Shuffle a deck of cards and place it facedown on the table. This is the draw pile.

② Name two cards—say, ace and king—to be the trigger cards. Turn cards over, one at a time, from the draw pile, and place on a discard pile. Only one card face should be visible at a time.

③ Whenever you draw a trigger card, try to name the card you turned over two turns ago. The face of this card will be hidden, so you will have to use your memory. Check to see if your guess was correct. If it was right, take this card and keep it in your hand. Then square up the discard pile and keep playing.

④ Your goal is to capture as many cards as you can. Eight cards is a perfect game.

⑤ When you get comfortable with the 2-back game, try one of these harder game variations:

- **Two decks**. For a longer game, add another deck of cards.
- **3-back.** Instead of recalling the card that appeared two turns ago, recall the card that appeared three turns ago.
- **2- or 3-back.** Whenever you turn over a 2 or a 3, remember the card that appeared that many cards ago.

P

N-BACK GAME WITH VOICE RECORDER *(Memory)*

▶ *This puzzle strengthens your working memory.*

1 You will need a sound-recording device or computer program.

2 Read the following sequence of letters aloud into a voice recorder or other recording device at a rate of about one letter per second.

```
PLAYF ULBRA INBIN UAPYL RBAPP YNFAR LYBUF
LFIUR NIINB NRAFP ULRAI LYRFP ULYYU AFINB
PBYAY PRFBA ULAII BLNPR LINFU FBYNP URUNY
LINUY FBIFB PRBAU FALLP INRPY RAPAF FIPAL
UBFYP LNUNR LYIYI BRANB UR
```

3 After a short break, during which you can occupy yourself with something else, play back the recording while listening for a randomly selected target letter—B, for instance. When you hear the target letter, stop the recording and write down the letter that you heard two letters previously.

4 Pick a new target letter. When the target letter comes up, repeat the process of writing down the letter that came up two letters before.

5 Check your answers on page 55.

Your brain uses several mental operations to succeed at the 2-back challenge. When you hear the target letter, you must stop the recording and write down the letter that you heard two letters back: the 2-back letter. In order to do that, you must constantly keep track of the last two letters heard. As each new letter is heard, a new 1-back letter must be stored and the old 1-back letter moved to the 2-back position. Finally, when hearing all other letters but the target letter, you take no action but keep listening for the target letter.

After you become skilled at 2-back, try 3-back: identify the letter you read three letters previously. In fact, you can go back as far as you want. That's why this class of tests is referred to as N-back: because *N* can be any number. Anything beyond 2-back is quite a bit harder, as you will find out. Don't be discouraged if you have to keep working for several weeks in order to become skilled in 3-back challenges.

As you improve your performance on the N-back puzzles, you will be challenging the prefrontal areas of the brain that are activated in working-memory tasks and, in the process, increasing your intelligence. Following are a series of related challenges. Try doing them in your head; only if you can't do that should you write the words down and rearrange them on paper.

MENTAL JUGGLING *(Memory)*

ANSWERS ON PAGES 55–56

▶ *These puzzles strengthen both your short- and long-term memory.*

Memorize one of the lists below. Take as long as you need. Now, without looking at the list, try the following challenges:

- Recite the list from memory. Check your answer.
- Recite the list backward. Check your answer.
- Recite the list in alphabetical order.
- Recite the list in order from shortest to longest. The words in each list all have different numbers of letters.
- Try the same challenges for each of the other lists.
- Bonus question for states: Recite states from west to east.

COLORS	NAMES
Red	John
Orange	Elizabeth
Goldenrod	Carol
Green	Daniel
Blue	Shirley
Crimson	Bob

RANDOM WORDS	STATES
Match	Florida
Picture	Utah
Ace	California
Upstairs	Hawaii
Base	Mississippi
Expert	Texas

▶ ▶ ▶

▶ ▶ ▶

REFLECTION. Which challenge was hardest? Which list was hardest? You probably found memorizing and reciting the lists a lot easier than the working-memory tasks in which you had to recite them backward or list them alphabetically or according to word length. In these exercises of your working memory, you had to perform mental manipulations in order to mentally rearrange the lists.

ADVANCED MENTAL JUGGLING (*Memory*)

ANSWERS ON PAGE 56

▶ *This puzzle strengthens both short- and long-term memory. The memory challenge is harder because you must construct a list entirely in your imagination.*

NUMBERS	OPPOSITES
1	Awake
3	Day
5	Fast
11	Inward
13	Off
15	Old

Numbers. Take a look at the list above labeled NUMBERS. Imagine the spelled-out name of each number. For instance, the first number spelled out is the three-letter word ONE. Once you have all the spelled-out names in mind, try the following challenges:

• Recite the number names in order from memory. Check your answer.

▶ ▶ ▶

N

▶ ▶ ▶

- Recite the number names in backward order. Check your answer.
- Recite the number names in order from shortest word to longest. Note that all the number names have different numbers of letters. Answer on page 56.

- Recite the number names in alphabetical order. Answer on page 56.

Opposites. Take a look at the list on the previous page labeled OPPOSITES. For each word on the list, think of the word that means the opposite. For instance, the opposite of AWAKE is ASLEEP. Once you have all the opposite words in mind, try the following challenges:

- Recite the opposite words in order from memory. Check your answer.
- Recite the opposite words in backward order. Check your answer.
- Recite the opposite words in order from shortest word to longest. Note that all the opposite words have different numbers of letters. Answer on page 56.

- Recite the opposite words in alphabetical order. Answer on page 56.

REFLECTION. How did you hold the secondary list in your memory? Did you visualize it? Did you rehearse it out loud?

Here are the 2-back letter sequences for each possible trigger letter.

P is the target letter: U B A A R N A L Y F L N R F F

L is the target letter: F P A U P A PA I P N F A P Y N

A is the target letter: P B N R N N L Y B F U R U Y A I B

Y is the target letter: L A P R I U L P Y F U N R B R Y

F is the target letter: A Y B F R Y U P I F U B A P A U

U is the target letter: Y I Y F F F Y B N N U I B A L N

B is the target letter: U I L L I I B R I U Y I P L Y A

R is the target letter: L Y F I B U L Y N P B I P U I B

I is the target letter: R N L R N R A L A R Y F L F L I

N is the target letter: A B P U I N F B L B R L P P N R

ANSWERS TO MENTAL JUGGLING

Alphabetized. Here are the words in alphabetical order.

COLORS	NAMES
Blue	Bob
Crimson	Carol
Goldenrod	Daniel
Green	Elizabeth
Orange	John
Red	Shirley

RANDOM WORDS	STATES
Ace	California
Base	Florida
Expert	Hawaii
Match	Mississippi
Picture	Texas
Upstairs	Utah

F

Short to long. Here are the words ordered from shortest to longest.

COLORS	NAMES
Red	Bob
Blue	John
Green	Carol
Orange	Daniel
Crimson	Shirley
Goldenrod	Elizabeth

RANDOM WORDS	STATES
Ace	Utah
Base	Texas
Match	Hawaii
Expert	Florida
Picture	California
Upstairs	Mississippi

ANSWERS TO ADVANCED MENTAL JUGGLING

Alphabetized. Here are the words in alphabetical order.

NUMBERS	OPPOSITES
11 Eleven	(Awake) Asleep
15 Fifteen	(Old) New
5 Five	(Day) Night
1 One	(Off) On
13 Thirteen	(Inward) Outward
3 Three	(Fast) Slow

Short to long. Here are the words ordered from shortest to longest.

NUMBERS	OPPOSITES
1 One	(Off) On
5 Five	(Old) New
3 Three	(Fast) Slow
11 Eleven	(Day) Night
15 Fifteen	(Awake) Asleep
13 Thirteen	(Inward) Outward

LONG-TERM MEMORY: IMAGINING THE FUTURE BY REMEMBERING THE PAST

Now THAT WE'VE worked on improving working memory, let's move on to long-term memory: the sum total of all the things stored in our brain from the past. This involves two qualitatively different forms of memory.

As an example of the first, answer this question: Who was the sixteenth president of the United States? In order to respond correctly to that question (the answer is Abraham Lincoln), you had to bring into your conscious mind a specific fact and then express that information in language, either spoken or written. Hence this form of memory is referred to as *declarative memory*. Among the examples of declarative memory are remembered telephone numbers, addresses, birthdays, and tax deadlines.

For most of these items of general information, it's unlikely you remember the circumstances in which you learned them. If you do remember, that declarative memory is referred to as *episodic*, i.e., you remember learning about Abraham Lincoln during an elementary school history class, perhaps even the events of that specific day. If you can't remember the exact time or specific details of how you first learned about Lincoln (the usual situation for most of us), that declarative memory is referred to as *semantic*. Most of our memories are semantic. In fact, for those few "gifted" individuals with predominantly episodic memories, the effect on their lives has been devastating.

Imagine reading a list of seventy words today and being able to recite it without error fifteen years later. The Russian psychologist Aleksandr Luria described a man named Shereshevski with such extraordinary recall. But rather than a benefit, his memory virtuosity was a torture to him. He remembered so much that he couldn't generalize or sort out the important from the trivial. Everything was recalled as a specific experience, and he was tortured with details that he neither needed nor desired. Imagine remembering every telephone number you have ever dialed. That will give you some idea of the burdensome world of Shereshevski, described in Luria's classic account *The Mind of a Mnemonist*. It's useful to keep Shereshevski's experience in mind whenever we wish for a superpower memory. As Oscar Wilde put it, "When the gods want to punish us, they answer our prayers." We don't want to remember everything that we've ever experienced but instead only a *select* portion of our experiences.

The second form of memory, *procedural memory*, involves skills and associations that rise to the conscious level only during a short period when we're first learning them. Riding a bike, playing a musical instrument, and swimming are examples of procedural memory. When first learning to swim, you concentrated on breathing, the optimal movement of your arms and legs, and so on. But as you became skilled, your swimming technique switched from declarative to procedural memory; in other words, you no longer paid conscious attention to your technique but simply dove into the water and effortlessly and unconsciously swam to the other side of the pool. Now if someone asks you to explain how to swim, you don't resort to explanations: you tell him to watch, and then you just dive into the water and start swimming. Your procedural memory takes precedence over your declarative memory.

Different brain areas are involved in declarative and procedural memory. During that history lesson many years ago when you first learned that Abraham Lincoln was the sixteenth president, that information was encoded in the hippocampus, the sea-horse-shaped structure in the temporal lobe that connects via an arching circuit, the fornix (Latin for "arch") to two other structures, the mammillary bodies and the dorsal thalamus. An additional structure, the basal forebrain—located, as the name implies, at the base of the anterior portion of the brain—is also linked to the fornix. Damage to any one of these three structures results in severe memory impairment. (On the positive side, you strengthen the circuits involving these structures whenever you exercise your memory. According to an elementary rule of brain functioning—"What fires together wires together"—the more you exercise these circuits by memory challenges, the stronger your memory becomes.)

After the Lincoln information was encoded within these memory-encoding structures, that information was widely distributed to the rest of the brain via association fibers. In general, long-term information storage occurs in select areas of the brain. Your recall of Lincoln's name involves language and vocabulary that is stored chiefly in the association area of the left temporal lobe. Your memory of Lincoln's bearded face is also stored in the temporal lobes. And if the technology had been available in Lincoln's lifetime to make a recording of his voice that you could listen to, your perception of that, too, would be stored in the temporal lobes. But different brain areas, the frontal lobes, would become activated if you were deciding whether Lincoln was an effective president. That kind of decision involves judgment and the weighing of alternatives. Think of your experience of Abraham Lincoln as involving many distinct brain modules working together to produce the totality of all of the things you have learned and can state (i.e., declare) about Lincoln.

Procedural memory involves entirely different parts of the brain. Included here are the basal ganglia, the cerebellum, and the parts of the motor cortex located toward the front of the brain (the premotor areas). You can think of the basal ganglia as separate clusters of cells located deep down beneath the cerebral hemispheres. These clusters communicate with the rest of the brain via association fibers. The basal ganglia control automatic processes that operate outside the sphere of consciousness. For instance, as your swimming skills improved, your basal ganglia took on a greater importance, ultimately displacing the association areas as the swimming coordinator. But if you later decide to engage in swimming competitions, the association fibers—especially those from the frontal lobes—will reestablish themselves. Declarative memory will once again become important as you absorb instruction from your coach and consciously apply the new knowledge gained from professional instruction as a means of improving your swimming performance. As you become more proficient, the shift from association fibers to the basal ganglia will occur once again. Declarative memory will be subsumed into procedural memory.

For the most part, we won't be dwelling too much in this book on procedural memory. But as you become increasingly skilled at solving puzzles, some of that skill will become part of your procedural memory, i.e., you will develop an intuitive sense for how to go about solving the puzzle. To get started, let's coordinate our knowledge about the brain mechanisms underlying memory with techniques you can use to improve your memory.

Retrieval and Learning

Study the following list for one minute and then repeat back as many of the items on the list as you can without error:

cag	suj	riy
los	fiv	lwx
moc	tij	gor
baf	coj	raj

Now do the same with this list:

cat	sun	rib
log	fin	lox
mob	tie	god
bat	cot	rat

Note that the words in the second list differ from the letter combinations in the first list by only one letter. So why did you find the second list easier to learn and remember than the first? Because the letters in the second list correspond to readily recognizable words, whereas the letters in the first list are arranged randomly into syllables that make no sense.

That change of one letter provides a wealth of meaningful and naturally occurring personal associations. For example, while everyone has some relationship with cats, whether positive or negative (they are cat lovers or haters, sufferers from cat allergies, fans of the musical *Cats*, etc.), *cag* is a meaningless combo of two consonants separated by a vowel. The greater ease in learning and subsequently remembering meaningful as opposed to meaningless material underlies the first law of memorization as developed by the world's first memory researcher, Hermann Ebbinghaus.

Ebbinghaus was a strange man who, during the latter part of the nineteenth century, wandered from one European university to another, earning a sparse living as a teacher and private tutor. One day, while browsing at an outdoor bookstall, Ebbinghaus found a textbook describing what was known at the time about human perception: briefly, people's perceptions of physical variations—the weight of a ball, the brightness of a light—can be expressed by simple mathematical equations. This inspired him to wonder whether memory might also be defined by mathematical laws. To find out, he devised a series of experiments. And because he was low on funds and couldn't hire experimental subjects, he was forced to carry out the experiments on himself.

Ebbinghaus created a list of 2,300 nonsense syllables, each consisting of two consonants separated by a vowel (*nog, baf*), similar to the first list you were asked to memorize. (Legend has it that Ebbinghaus took his inspiration for this from Lewis Carroll's *Through the Looking Glass,* which contains verses of rhyming nonsense words.) He then memorized a series of lists containing various numbers of these pseudo words and tabulated how long it took to learn them and how easily he forgot them.

From Ebbinghaus's somewhat bizarre research, published in 1885 in his monumental *Memory: A Contribution to Experimental Psychology*, the following principles emerged:

First, according to the human learning curve (Ebbinghaus's term), the time required to memorize a list of nonsense syllables increases as the number of syllables increases. Second, memorization of meaningful words takes only about one-tenth of the effort required to learn comparable nonsense material. Third, multiple learning trials distributed over time are more effective for memorizing than extended one-session efforts. Finally, continued practice after learning enhances retention.

While all four of these learning principles of memory remain valid today, another commonly accepted principle has turned out to be incorrect: the belief that testing does not produce learning but serves only as a means to measure how much has occurred. Not true, according to two psychologists from Purdue University writing in the prestigious journal *Science*.

"Testing is the critical factor for promoting long-term recall," authors Jeffrey Karpicke and Henry Roediger III maintain. Here is a brief reprise of their research:

Imagine yourself learning the Swahili equivalent of forty English words (e.g., *mashua*, or "boat"). Unless you speak or have some familiarity with Swahili, learning those equivalents is likely to take some time. But after you finally learn the words, Karpicke and Roediger found that additional studying of the words was less effective in enhancing long-term recall than repeated testing of the word pairs.

"The conventional wisdom expressed in many study guides is wrong," write Karpicke and Roediger. "Although educators often consider testing a neutral process that merely assesses the contents of memory, practicing retrieval during tests produces more learning than study alone once an item has been recalled. Even after items can be recalled from memory, eliminating those items from repeated retrieval practice greatly reduces long-term retention."

In other words, although educators typically consider testing a neutral process that merely measures how much has been learned, *testing actually results in greater learning.*

Think back to those ornery professors in college (only a small minority of teachers: that's why we remember them so well) who favored "pop quizzes." Maddeningly, these unpredictable and anxiety-arousing quizzes often included questions correctly answered on previous tests. ("We've already been tested on that, Professor. Why do we have to come up with those answers again?") Well, it turns out those professors were on to something about retrieval learning. You cannot simply learn something, never test your ability to retrieve it again, and then expect to retain that information over the long term.

The Brain's Memory Circuits

What explains this powerful effect of retrieval on learning and memory? It all has to do with the formation of neuronal circuits corresponding to the information learned. As noted earlier, the Canadian psychologist Donald Hebb proposed in 1949 that the brain stores memories in the form of networks of neurons that he referred to as cell assemblies. These networks are strengthened each time the memory is retrieved. He also suggested that learning involves the strengthening of these networks on the basis of our personal experience. In one of the most quoted passages in the history of neuroscience Hebb wrote: "When an axon of cell A is near enough to excite cell B and repeatedly or persistently takes part in firing it, some growth process or metabolic change takes place such that A's efficiency, as one of the cells firing B, is increased." In other words, the strength of the connections between neurons gets stronger as the neurons become simultaneously active.

Each time you activate one cell in a cell assembly, it makes it easier and more likely for the other cells to fire as well. Thus knowledge, habit, and skill formation involve, first, the formation of cell assemblies and, second, the strengthening of these assemblies through frequent stimulation. If the cell assemblies are weakened through infrequent use, they will eventually be disassembled, resulting in the loss of knowledge or skill formation. "Use it or lose it" is the popular and accurate mantra describing this process.

A second principle is also at work whenever we activate a cell assembly. A memory doesn't exist in the brain like a read-only file on a computer. Instead, memories are dynamic and must be reconsolidated each time we remember them. Biochemical research shows that each time we call something to mind, our brain synthesizes additional proteins, which is similar to what happens when we first establish new memories. In other words, each time we remember something, our brain creates a newly reconsolidated version of it, resulting in multiple traces of the same memory. This is both good and bad. On the positive side, we can call upon any one of several traces and thus strengthen the memory, as with the "pop quiz" example mentioned on page 63. On the negative side, careless recall can lead to false memories. This is one explanation why our memories for the same event change over time. It's also why other people's suggestions can exert a powerful effect on changing our memories. Numerous experiments carried out at the University of California, Irvine, by psychologist Elizabeth Loftus and others show that a person's memory can be altered by providing him with false information in the form of a question. For example, if a witness to a car accident at an intersection controlled by a signal light is asked, "Just prior to the accident, which car failed to stop at the stop sign?" that misleading information will create a false memory of the circumstances of the accident. Because human memory is so susceptible to modification, judges routinely admonish attorneys for asking "leading" questions, or questions that subtly introduce the desired answer and, as a consequence, alter witnesses' memories.

But such attempts at memory deception aside, repeated testing leads to the creation of multiple memory traces, which aid in subsequent recall. Returning to the Swahili–English research mentioned earlier, once the word pairs are learned, the links are additionally strengthened each time the word pairs are retrieved. The process is no different from what occurs whenever we learn a physical skill such as tennis or golf. Once we have developed a finely honed backhand or a deadly-accurate putt, we have to keep practicing it in order to maintain it. Practice in the physical and mental sphere is another word for retrieval.

This process of learning through repetitive retrieval is something we've been practicing our entire lives. For instance, as infants and children we learned to speak our native language by repeating many of the basic forms again and again. We still do. (Count how many times you say "How are you?" or "How are things going?" on a typical day.) Lack of retrieval accounts for many memory failures. Our inability to come up with the names for people, places, and specific items of information often occurs because we haven't thought about (i.e., retrieved) that particular piece of information for a long time.

Here's an example: Because I grew up near a Civil War battlefield, I studied the Civil War very intensely during my grade school and high school years. But as my interests and career choices over the years have taken me in a different direction from military history, I now have a bit more trouble coming up with the names of generals, battles, the total number of troops in specific engagements, and other particulars. Yet only a few minutes spent in a quick review of these facts, coupled with sitting back and testing my ability to mentally retrieve them, refresh my memory. It isn't that I have forgotten those Civil War details. Rather, I experience difficulty bringing them to mind because of my failure to retrieve them frequently enough over the years to keep the relevant brain circuits sufficiently strengthened.

The Link Between Remembering and Doing

Every time you remember something, you exercise neurons in the same brain circuits that were first established during the original experience. Proof for this comes from research carried out on epilepsy patients undergoing a presurgery workup directed at pinpointing the sources of their seizures. Before surgery, the researchers inserted electrodes into about one hundred cells around the memory-encoding hippocampus and measured electrical activity from the cells as the patients watched short video clips depicting landmarks (like the Statue of Liberty), characters from popular television shows (Homer Simpson), or animals (cats). Many of the cells responded to at least one of the clips.

Later the patients were asked to remember the clips. As they brought one or another of the clips to mind, the same cells that had fired during the original viewing of the clips fired once again. This pattern was sufficiently reliable that the researchers could even predict which clip the patient was remembering a second or two before he described it. Put another way, the same neuron that preferred a particular clip became excited both when the patient looked at the clip and when he later remembered it.

A similar remembering-doing link exists when it comes to carrying out actions and then later repeating them. For example, if a recording is made of a trained rat's hippocampal neurons while it runs through a maze in a specific pattern (alternately entering the right versus the left arm of the maze), it soon becomes possible for an observer to predict which path the rat will take by simply studying the rat's hippocampal firing patterns. Those patterns, originally established when the rat first learned to maneuver the maze, repeat themselves each time the rat "chooses" to run either the right or the left arm. If it's time to enter the right arm of the maze, the same neurons that fired the last time the rat "chose" the right arm will fire this time as well. If the rat makes a mistake and runs the wrong arm of the maze (left instead of right), that error can be predicted beforehand by observing the firing patterns immediately preceding the rat's "choice" of path, which, because of this error, will signal the left rather than the right path.

We're essentially talking here about the power of the brain to establish memory circuits, maintain those circuits, and reactivate them upon demand. One proviso, however: Memory is like a muscle and must be exercised in order for it to remain strong. Before we suggest how to do that, let's consider the role of memory as forming a bridge linking past, present, and future.

Imagining the Future by Remembering the Past

Each time we exercise our memory, we not only enrich our grasp of the past but also strengthen our ability to envision the future. At first this sounds paradoxical: traditionally speaking, both past and present can be spoken of as "real." The future, however, doesn't exist—except in science fiction novels.

Quickly think of a party you attended in the recent past. Mentally zoom in on five of the people who were there. Try to picture them as clearly as you can: how they were dressed, your interactions with them, perhaps something special that occurred at the party involving one or more of them. Got it? Now clear your mind and imagine a future party held at a different time and for a different purpose. Imagine it taking place ten years from now and with those five people (along with yourself) a decade older. You, along with one or more of those people, may now be married or divorced; children may be present or may have grown up and moved away, depending on everyone's age at the time of the first party; some of the partygoers may be in new jobs or may have retired—again, depending on everybody's age at the time of the first party. It's easy to imagine many other changes in your imagined party. Would you be surprised, as I was, to learn that the same brain circuits are involved in these two seemingly very different exercises—that your ability, in this example, to imagine the future party vitally depends on the clarity of your memory for the past party?

Over the years, psychologists have observed that people experiencing memory failures also aren't good at imagining future experiences. The first hints of this emerged from the study of brain-injured patients. After a serious motorcycle accident in which he suffered widespread brain damage, a patient referred to by his psychologist as KC could neither remember anything from his past nor imaginatively project himself into the future. In response to requests to remember or imagine personal events, KC experienced mental "blankness." Similar failures occur in people suffering from depression. Among people between twenty and forty years of age who complain to their neurologists of declining memory, depression is a frequent cause. Such patients experience difficulties in imagining themselves ever feeling better—one of the reasons depression is difficult to treat. But we don't have to posit brain damage or depression to encounter impairments in remembering the past and imagining the future: similar difficulties occur in the perfectly normal brain.

Think about your friends and colleagues. You probably know of one or more with excellent memories; others who depend heavily on memory-jogging techniques like day journals and the calendars available on BlackBerrys and other electronic devices; and finally, the large majority of the group, whose memory abilities exist somewhere between these two extremes. Similar differences exist among people when it comes to imagining the future.

Again, think of your friends and colleagues. Some are virtuosos when it comes to imagining the future consequences of taking or not taking certain actions, such as "telling off" the boss in response to what seems like an unreasonable request versus doing what's requested and getting on to less irksome aspects of the job. Others are seemingly unable to put current difficulties into perspective by adopting what Buddhists refer to as the "long view." Most of us dwell somewhere in between these extremes.

What takes place in the brain when we imagine the future and remember the past? Several of the same regions toward the posterior of the brain are activated during both processes: the occipital cortex, the posterior cingulate cortex, and the medial temporal cortex. These areas become active when we respond to an autobiographical question such as "Name two of your best friends in college" or when we mentally imagine ourselves cruising through our neighborhood at the wheel of a bright red Ferrari. Why would these same areas become active when remembering something from our past and fantasizing something happening in our future? It's because we tend to imagine future events in terms of people and places that are already familiar to us, even though we don't do it in exactly the same way.

In the party example, you're likely to envision the future party by incorporating elements from the original party: perhaps the same location, similar weather (if the original party was held outdoors, etc.). Remembering is like scanning a mental photograph; imagining is like creating a collage composed of snippets from many photographs arranged in novel patterns. Since imagining is a more creative and demanding task, additional brain areas are called into play. These include regions involved in imagining a behavioral sequence such as returning a tennis serve or reaching the halfway point of the daily jog. What's the take-home message in all of this? Those brain regions known to be important to our memory of our past also play an important role in helping us to imagine our future. This has practical implications: Whenever you work on improving your memory, you simultaneously contribute to your imaginative and creative capacities.

Specific Techniques for Boosting Your Memory Power

Recall that the working-memory puzzles you tried earlier involved both visual-spatial and verbal challenges: some puzzles required you to manipulate structures in mental space while others challenged your facility with manipulating words. Each of these two different functions can be used to bolster the performance of the other. For example, if you want to remember the names of several people you've just been introduced to at a cocktail party, silently enunciate the names while converting them into pictures based on the people's first or last names, their appearance, or some other distinguishing visual characteristic. When you do that, you combine visual-spatial and verbal working-memory systems to aid your later recall of those names. Try the same approach for remembering the name of someone you've read about. For instance, I was recently researching Sigmund Freud's nephew, Edward Bernays, the author of the book *Propaganda* and one of the founders of the public relations industry. As an aid in remembering his name I visualize béarnaise sauce being poured over a succulent steak.

But the value of combining these approaches isn't limited to names. Any word or complicated concept can be replaced with an image. In general, it's easier to remember if you create bizarre images in which the items to be remembered play prominent roles. For instance, the words *key, car, mortgage,* and *scissors* can be recalled via the image of a car outfitted with a steering column shaped like a key with scissors instead of tires being driven over a huge sheet of official-looking paper. As the car advances, the scissors open and close, shredding the paper into ribbons. The clearer that image is envisioned and the more often it is retrieved, the easier it is to remember those four words.

Another useful visual-spatial procedure involves placing the items to be remembered in mental space. For instance, a waiter in a nearby restaurant never writes down his customers' orders, even when serving large tables. He learned to perform this seemingly formidable memory exercise by first memorizing the menu. As each customer places his order, the waiter mentally substitutes that item on the menu with an internal picture of the customer. Later, in the kitchen, he places the order by reconverting from the picture to the menu item. If he forgets anything, he has only to immediately see in his mind's eye the various customers arranged according to their places on the memorized menu. In response to requests for variations on the menu items, he pictures the requested variation and then highlights an image of it beneath his mental picture of the customer's selection.

Another memory technique includes memorizing acronyms such as ROY G. BIV for the colors of the visible spectrum (red, orange, yellow, green, blue, indigo, violet). Medical students remember the twelve cranial nerves (olfactory, optic, oculomotor, trochlear, trigeminal, abducens, facial, auditory-vestibular, glossopharyngeal, vagus, accessory, hypoglossal) by silently reciting the jingle "On Old Olympic's Towering Tops A Finn And German Vied At Hops."

Another system, the *method of loci*, involves memorizing a list of objects by mentally envisioning a familiar place, such as one's living room. The items to be remembered are then distributed at familiar locations. They can later be recalled by simply mentally strolling through the living room and observing the locations of the objects you want to remember. But if you want to permanently maintain the information, your mental stroll must be repeated frequently in order to strengthen the relevant circuits. Think of the items as associated in a network, just like neurons linked together in a circuit that grows stronger the more frequently the circuit is activated.

When remembering words, also take advantage of the *phonological similarity effect*: your working memory for words is better when the letters sound different (e.g., S, R, and X) than when they sound similar to each other (e.g., P, B, and V). This effect suggests that the brain stores information in a phonological format, i.e., according to the patterning and rules governing how words are pronounced. The more unusual the word and the more novel its pronunciation, the easier it is to remember.

Remembering words is also influenced by the *word-length effect*: you remember more words when the words are short. So the shorter you can make the words, the more of them you will be able to hold in your working memory.

Words are also easier to learn when they rhyme (the reason poetry is easier to memorize than prose). But be certain not to use words that sound too similar to each other because you may inadvertently overstress the left inferior parietal cortex, which is responsible for the phonological similarity effect.

Finally, it's easier to remember something if you break it down into smaller units, such as dividing seven-digit telephone numbers into three- and four-digit segments. As a corollary, something is more easily memorized if it can be "chunked" into meaningful segments such as IBM2008AAANFL remembered as IBM 2008 AAA, and NFL.

Memory establishment also varies according to the category that you're trying to memorize. Neuroscientists have known for a long time from fMRI brain scans of volunteers that viewing pictures of houses or furniture leads to greater activity in the medial portion of the fusiform area, located in the temporal lobe. Looking at pictures of faces, in contrast, yields greater activity in the lateral fusiform areas. Variations in the functioning of the fusiform face area no doubt play a part in the variations among different people in their facial recognition abilities. Some of us possess a seemingly uncanny talent for recognizing faces and linking them up with their owners' names; others experience great difficulty recognizing even familiar faces when encountered in unfamiliar surroundings.

Things That Interfere with Memory

Interruptions of attention are the most frequent cause of memory failures. If you aren't really listening when you're introduced to someone at that cocktail party, you won't remember his name later. As a result of your inattention, your brain never had the opportunity to register and encode the name within the hippocampus, the first link in the memory chain. Absent registration and encoding, the brain wasn't able to transmit the name along the association fibers for eventual storage within the cerebral hemispheres. Any break in the chain involving encoding, transmission, and storage results in memory failure.

Our level of alertness plays a big part in our memory. If we're either too drowsy or too "hyper," we can't focus our attention effectively on the material we want to learn and therefore won't remember it later.

According to the Yerkes-Dodson law, arousal and attention are related as an inverted U-shaped curve, with optimal attention occurring at midpoint in the curve: not too tired, not too wired.

But even when we're fully alert and paying attention, irrelevant information can create conflict in ways that are sometimes difficult to circumvent. For example, when color names are written in non-matching ink (e.g., the word *green* is printed in red, the word *orange* is printed in yellow, etc.), deliberate effort is required to avoid reading aloud the color word if we have been requested to name instead the ink color that the word is printed in. That's because over a lifetime of experience we've become accustomed to automatically reading written words that we encounter as words in our native language. Of course, it's *possible* to say aloud the color names instead of the word in this so-called Stroop test (named after its developer, John Ridley Stroop), but it takes longer because of the deliberate effort we have to exert to inhibit our "natural" tendency to read the word. To experience this effect for yourself, have a friend write out a list of color names using ink of different colors along with a second list in which the color names and the color inks match. Now time your speed at reading both lists. Even though you're aware of the "gimmick" involved in this exercise, you will need more time to process the word–color mismatch list.

In a variant of this test, words with stronger emotional connotations (*murder, rape, execution*) take longer to read than neutral words (*furniture, necktie*). This so-called emotional Stroop effect results from the fact that emotional words capture our attention, momentarily distract us, and slow our response time. The effect also varies according to personality types and specific problems: drug-related words produce a delayed response for someone with a drug problem; phobia-related words (*web, tarantula*) slow the Stroop response for someone with a fear of spiders.

Identification of the brain activity occurring during the processing of emotional words can be demonstrated by what psychologists refer to as masking experiments. Here a word is shown for just a few tens of milliseconds, immediately followed by an image of an object or a face, which prevents conscious recognition of the word. As the delay between the word and the image increases, the word gradually becomes recognizable. Usually the masked word suddenly "pops" into consciousness when the interval reaches about 50 milliseconds. But if the word is an emotional one, recognition occurs more slowly: the amygdala responds to events and (in the emotional Stroop test) to words that are emotionally loaded. For a few brief milliseconds, it preempts the process of reading words.

Nor are these effects limited to the neurophysiologist's laboratory; they also occur in everyday life. For instance, it's difficult to ignore something insulting said about you in your native language. In fact, addressing someone in an insulting way in a language that the person claims not to understand but actually does has been used in wartime as a means of uncovering spies. Exempting the most skilled practitioners of the art of emotional concealment, hearing and understanding oneself described in demeaning terms elicits an automatic response (some outward sign of anger) in anyone who understands what has been said. Our name has an especially powerful effect on copping our attention because of the effort required to block out the normal alerting response that occurs when we hear our name called. For example, when my parrot, Toby, calls out "Richard!" I find it difficult to retain my concentration; and after the bird has repeated it a few times, I become sufficiently distracted that I'm forced to (1) stop what I'm doing at the time, (2) give Toby some attention, or (3) move him to a cage in another room.

Attention depends on the normal functioning of a network of neurons located principally in the parietal lobes, with some contributions from the frontal lobes. Damage to either of the parietal lobes creates serious difficulty in attending to anything on the opposite half of the body, including things seen on that side (the visual field, as neurologists refer to it). But there is an important difference between the right and left parietal lobes. While the left parietal lobe attends primarily to the right side of the body, the right parietal lobe attends to both sides. This makes damage to the right parietal lobe especially impairing since the left parietal lobe cannot compensate for loss of attention to the left side. This discovery from brain-damaged patients has practical consequences for the optimal function of the normal brain.

In general, your chances of remembering a map or diagram will be greater if the image is stored in both rather than only one parietal lobe. So if you want to remember something of a nonverbal nature (a map or a diagram, for example), place it off to your right side. This will activate both parietal lobes and thereby encode the information in both sides of the brain. This will be more helpful for remembering than if you place the map or diagram to your left side, which will activate only the right parietal lobe and, as a result, the information will be stored only on the right side of the brain. As another application of this insight, select a seat on the left side of a classroom facing the teacher during classes involving a lot of images, such as art appreciation or architecture.

Additional Memory Enhancers

Think as deeply as possible about what you're trying to memorize. For example, learn the word's origin; break the word down into its components; form a mental picture of the word. Mentally picturing the word brings more brain structures into play than a more superficial approach, like silently spelling the word in your mind. In one fMRI study illustrating this distinction, brain activity was especially high within those areas known to be associated with memory, such as the hippocampus and nearby areas of the medial temporal lobe. "This suggests that the brain is working harder when forming an image," according to memory expert Mark Gluck. "Deep versus superficial processing seems to correspond to measurements of how hard the brain is working."

Forcing the brain to work harder increases the chances for later recall, according to fMRI studies. In one especially compelling experiment, fMRI images were taken while volunteers looked at a list of words to be memorized. Later, the fMRI images were correlated with the words that could be remembered and the words that had been forgotten. Greater activity occurred in the medial temporal lobe for words that could later be remembered. The greater the activity, the more efficient the storage and, as a result, the greater the likelihood that the subject would remember the word.

Finally, memory works best if you duplicate as closely as possible the circumstances existing at the time you tried to memorize something. In one rather bizarre experiment illustrating this principle, members of a diving club learned a list of forty words either while underwater or while on land. Both groups remembered more words when they were tested later under the same conditions that they had originally studied the words. Later that year, the same team of researchers turned their attention away from divers and concentrated on students who learned a word list while either standing or sitting. Both groups recalled more words if they were tested while in the position they occupied when they originally studied the words.

Even background music can influence memory. If you listen to jazz or classical selections while memorizing, you'll remember better if you listen to the same type of music when recalling the memorized material.

All of these memory aids work because they facilitate the formation and strengthening of neuronal circuits. And each time we activate these circuits by using one of these memory aids, we solidify what we've learned. Furthermore, circuits don't exist in isolation but interact, as with the cat example given on page 61: every time we think of a cat, we activate circuits related to our feelings about cats, allergies, Broadway plays, and a host of personal experiences involving felines. Philosophers have referred to this integration and linking as the "unity of knowledge," but whatever you call it, an important measure of our mental capacity is our ability to activate these neuronal memory circuits.

The following games will help you develop different techniques for remembering things. Try these games and see which memory techniques work best for you.

MEMORIZING WORDS (*Memory*)

▶ *This puzzle strengthens your long-term memory.*

Memorize the following list of ten random objects:

> chair
> cat
> pencil
> umbrella
> ace of spades
> steering wheel
> sandwich
> hat
> rake
> harmonica

Try the following techniques:

1 Practice. Read the words in order; practice till you have them all.

2 Story. Tell a story that incorporates all the objects in order.

3 Picture. Make a mental picture in which each object is stacked on or otherwise attached to the previous object. For instance, start by putting the cat on the chair, and then put the pencil in the cat's mouth.

4 Imaginary journey. Imagine walking from the front of your house to the back, placing the objects in memorable places along the way. For instance, you could start by placing a chair in the front doorway. Ancient Roman orators used this technique to remember long speeches.

REFLECTION. Which technique worked best for you? Which technique allows you to recite the list backward?

MEMORIZING NUMBERS *(Memory)*

▶ *This puzzle strengthens your long-term memory.*

Here are the first 32 digits of the mathematical number pi:

3.14159265358979323846264433832795

Memorize them! Try the following techniques:

❶ Practice. Just start from the beginning and try remembering the sequence. It's better if you can speak the sequence to someone who is holding this book.

❷ Chunking. Memorize the same 32 digits in groups of four. Telephone numbers and credit card numbers are broken into chunks of three or four digits to make them easier to read and remember.

3141-5926-5358-9793-2384-6264-3383-2795

❸ Song. Sing the digits to a familiar tune, such as "Twinkle, Twinkle, Little Star." (You won't quite make it through the whole melody.)

❹ Mnemonic poem. The number of letters in each word of this poem gives a digit of pi. SIR = 3, I = 1, SEND = 4, etc. This poem gives the first 31 digits; add a 5 at the end to get 32 digits.

Sir, I send a rhyme excelling,
In sacred truth and rigid spelling,
Numerical sprites elucidate,
For me the lexicon's dull weight,
If nature gain, not you complain
Tho' Dr Johnson fulminate.

▶ ▶ ▶

REFLECTION. Which memory technique worked best for you? Try using these techniques to memorize important numbers, for instance, your Social Security or credit card numbers.

By the way, the world of pi memorization is full of astounding tales. The world-record holder for memorizing pi is 100,000 digits, set by Akira Haraguchi of Japan on October 3–4, 2006. It took him sixteen hours over two days just to recite all the digits. To accomplish the task, he used a mnemonic system that associates numbers with letters, enabling him to turn pi into a story.

"Cadaeic Cadenza," a 1996 short story by Michael Keith, is written so the length of each word gives the next digit of pi. The story is nearly 4,000 words, and encodes the first 3,834 digits of pi (some words encode more than one digit). The invented word *cadaeic* itself encodes the first few digits of pi: *c* is letter 3 of the alphabet, *a* is letter 1 of the alphabet, *d* is letter 4 of the alphabet, and so on. Not content with a short story, Keith went on to write an entire book, *Not a Wake*, encoding the first 10,000 digits of pi.

R

MEMORIZING NAMES AND FACES *(Memory)*

▶ *This puzzle strengthens your long-term memory.*

❶ Get a pencil and paper. You'll need them in a moment. Here are twelve faces and names. Take a couple minutes to memorize which name goes with which face. Then cover up the faces and names with a magazine or sheet of paper and go on to the next page.

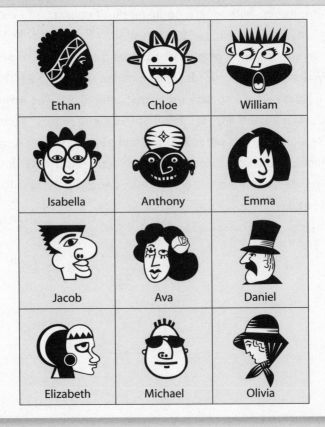

Ethan	Chloe	William
Isabella	Anthony	Emma
Jacob	Ava	Daniel
Elizabeth	Michael	Olivia

▶ ▶ ▶

▶ ▶ ▶

2 Here are the same twelve faces in a different order. Write down on your paper the names that go with the faces. Then check your answers on the next page.

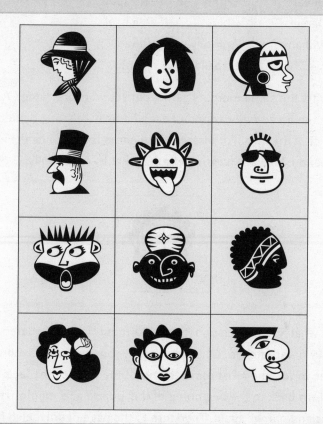

▶ ▶ ▶

▶ ▶ ▶

3 Here are the names that go with the faces on the previous page. How many did you get right?

Olivia	Emma	Elizabeth
Daniel	Chloe	Michael
William	Anthony	Ethan
Ava	Isabella	Jacob

Now try the same memory game, but with a new strategy. Associate the meaning or sound of each name with some distinctive aspect of the face. For instance, the name Isabella has the word *bell*, and Isabella is wearing earrings that look like bells.

Isabella

Use whatever association comes to mind first, no matter how quirky or far-fetched. Chances are that your first association will come to you the most easily the next time you see the face.

Turn back to the beginning of this puzzle and memorize the faces and names again. Then turn to the page of unlabeled faces and see how many names you can remember.

REFLECTION. Does the new memory strategy improve your performance? Try using this technique when you meet someone new, and see if it improves your memory for names.

The first step toward improving memory involves learning to focus attention despite distractions. As with the forgotten-name-at-a-cocktail-party situation described earlier in this chapter, you cannot encode and later retrieve unless you were originally paying attention. The following puzzles challenge your ability to override the brain's tendency to become distracted and latch on to incorrect interpretations, even when you've been specifically instructed to guard against making unwarranted assumptions.

MEMORY INTERFERENCE *(Memory)*

▶ *This puzzle challenges your long-term memory.*

Copy one of the following lists by writing it on a piece of paper or typing it on a computer. Read the words aloud and put the paper away; after five minutes, write down as many words as you can remember. Then check your new list against the original list. Which words did you miss? Which words did you add? (Warning: I've written each list to try to trick you into remembering a word that is not on the list.)

LIST 1

bed	rest
tired	snore
dream	nap
snooze	blanket
pillow	pajamas
robe	sheet

▶ ▶ ▶

▶ ▶ ▶

LIST 2

hungry	eat
spoon	bowl
cereal	morning
eggs	bacon
juice	coffee
newspaper	toast

LIST 3

headlight	tires
seatbelt	bumper
trunk	gas
key	brake
park	windshield
steer	turn

LIST 4

waves	sand
surf	ocean
tide	umbrella
chair	shovel
sun	pail
swim	lifeguard

REFLECTION. What techniques did you use to memorize the words in the list? What do you notice about the words you omitted or added? Were they related to other words in the list?

Perception

3

PERCEPTUAL SKILL LEARNING: THE
SOMMELIER AND THE HOCKEY PLAYER

WHAT DO a neuroradiologist, a jeweler, and an art appraiser have in common? Over time and after much practice, each has developed a finely honed talent for evaluating, respectively, brain-imaging studies, gems, and works of art. Initially, the learning process involved memorization: that radiologist would never have developed her ability to read CAT scans or MRI scans if she hadn't first learned brain anatomy. But memorization wasn't enough. True expertise required her to learn how to recognize the sometimes subtle neuroimaging differences that distinguish a normal brain from a diseased one. In order to do that, she had to review hundreds, even thousands, of imaging studies.

Perceptual skill learning is the neuropsychological term for the process whereby, through practice, experts encode in their brains the relevant perceptual skills needed to detect the first MRI suggestions of a tumor, appreciate the fine quality of a diamond, or distinguish a genuine Rembrandt from a fake. Once this is established, much of their subsequent work takes little time and seems effortless.

But perceptual skill learning isn't confined to specialized professional pursuits: it underlies such everyday challenges as understanding spoken and written communication in our native language. Memorization plays little role here; for the most part, we speak and understand our native language automatically (exempting vocabulary weaknesses). Becoming skilled in a new language, in contrast, depends very much on memory: starting from scratch, we have to build up a vocabulary, learn grammatical constructions, etc. But after sufficient practice, we may become almost as skilled in the new language as in our native language. Thanks to progress in our perceptual skill learning, we eventually recognize and employ formerly foreign words almost effortlessly. When we achieve this breakthrough, we no longer have to consciously rework or reexamine what we intend to say in the new language: we *just say it.*

Other examples of this shift from perception to action include parallel parking on a crowded street, working the clutch and shifting the gears of a manual transmission, and mastering a new guitar melody. Each of these examples involves a shift in emphasis from brain areas devoted to perception and attention (primarily the parietal lobe) to motor activity (the motor and sensory cortices). Practice is the key to bringing this about. Although learning to drive, for instance, initially requires effort and full concentration, we eventually reach the point where we can drive while conversing, listening to the radio, or even (don't do this) talking on a cell phone. As discussed earlier, if you practice something long enough and intensely enough, procedural memory eventually takes over from declarative memory. You no longer mentally rehearse what you're doing, you *just do it.*

As another example of perceptual skill learning, do you consider yourself a wine connoisseur—good enough to distinguish the powerful, tannic, and intense Grand Vin de Château Latour from the lighter silk-and-lace elegance of a Musigny from Domaine Comte Georges de Vogüé? How about perfumes? Can you distinguish Shalimar, with its strong Orient-inspired notes of bergamot and vanilla, from Chanel No. 5, with its sandalwood and amber notes added to the bergamot and vanilla? If you'd be uncertain or uncomfortable making such distinctions—and most of us would—you probably would come up with an excuse along these lines: "I don't have any special talents for distinguishing tastes or scents because my nose and taste buds aren't sufficiently sensitive."

Actually, wine and perfume expertise depends not on the nose or mouth but the brain. "The average person probably detects odors at about the same concentration as the professional wine taster or perfumer," according to Avery Gilbert, a sensory psychologist and consultant to the fragrance industry. "The expert has cognitive skills that make better use of the same sensory information."

Brain imaging supports Gilbert's claim. In one study ("The Appreciation of Wine by Sommeliers: A Functional Magnetic Resonance Study of Sensory Integration"), the brain activation patterns of wine experts were compared with those of people who had no special expertise in wines. As each sipped and savored wine samples, only the sommeliers showed brain activation in the orbitofrontal cortex, an area used when making judgments or decisions. A similar pattern occurred when testing the finer points of perfume appreciation. Greater orbitofrontal activation occurred among professional perfume experts compared with people who had no special training or interest in perfumes. The latter showed activation restricted to the primary sensory areas and regions associated with emotions. Such findings suggest that the experts develop their proficiency through the regular exercise of specialized mental rather than sensory skills.

In practical terms, we can develop what seems like extraordinary sensory acuity simply by honing our perceptual skills. When trying a new wine, for instance, follow the example of sommeliers: Think analytically, come up with words that describe what you are experiencing, and make notes for future comparison and reference.

But not all perceptual skills lend themselves to verbal description. A ballerina, for instance, cannot simply instruct beginners about what she does and how she does it. Her words cannot capture the complexity of the required motor-perceptual skills. Instead, the ballerina demonstrates the moves and the pupil practices them. When it comes to dancing, it is easier to show than to tell.

Once effortless performance is achieved on the basis of finely honed perceptual skill learning, conscious attention to the details of that performance often turns out to be not only unhelpful but outright counterproductive. This is especially true with amateur performers as opposed to experienced professionals. In an experiment demonstrating this, experienced golfers and amateurs with less experience putted while listening through earphones to a series of recorded tones. Each time the golfer heard a specified tone, he was asked to say the word *tone* out loud. Among experts, as opposed to amateurs, the tone-monitoring task did not decrease putting accuracy. In fact, among experts, the greater their accuracy in identifying the target tone, the more accurate their putting. Those with lesser degrees of putting proficiency, however, were less able to monitor the tone without undergoing deterioration in performance.

A similar effect occurs among hockey players. An experienced player has no difficulty skating and stick-handling a puck through a slalom course of pylons while simultaneously identifying geometric shapes projected onto a screen placed overhead. When novices try the same exercise, their skating and stick-handling abilities decline significantly.

Nor is this curious effect limited to sports. Skilled pianists have no problem sight-reading music while listening through earphones for a series of words. Amateur musicians find this distracting and as a result commit frequent mistakes.

What is the explanation for this? What is actually going on in the brain when procedural memory takes precedence over declarative memory?

At the early stages of learning a skill, various motor control centers in the brain must be integrated to achieve skilled execution. As a result, all attention must be directed to the integration of the separate circuits (movement, sensation, technique, etc.) underlying the controlled performance. Once this skill is mastered, attention is no longer needed for integrating the various components: the performance becomes automated. At this point, shifting the emphasis back to one aspect of the performance tends to prove temporarily detrimental to the performance as a whole.

As an example, think of the last time you concentrated consciously on the separate movements of your legs while you were dancing. After a few seconds of heightened attention to a segment of the complete sequence, your usual rhythmic smoothness fragmented into awkward stumbling. If you continued fragmenting your ordinarily integrated movements into separate overly attended-to components, you were soon falling all over yourself.

Typically this transition from a consciously directed to an automated response requires sustained effort and doesn't always follow a smooth trajectory. For instance, so-called beginner's luck in games and sports sometimes enables a novice to defeat a more experienced player. But such success lasts only a short time, because the motor programs necessary for consistent motor performance haven't yet been firmly established by the beginner. As a result, "beginner's luck" runs out when the beginner competes on multiple rather than single occasions against a more skilled and experienced competitor who, as a result of more experience, has established the relevant brain circuits.

Here's a somewhat pernicious practical application of this principle that you can use the next time you find yourself competing at tennis or golf with another amateur player who not only is performing better than you but also seems to do so effortlessly. Compliment him, and then, via suitably flattering remarks, ask him detailed questions about some aspect of his technique (his follow-through, the "exact" placement of his hands on the racket or golf club, etc.). This will refocus his attention however briefly from his integrated performance to a specific aspect of it. Such a shift in attention from the "big picture" toward details may give you a temporary edge, because it momentarily interrupts the automated aspects of your opponent's performance. Too much conscious attention to the separate components of an ordinarily automated performance also explains why a superior amateur player sometimes "chokes under pressure" and suffers defeat at the hands of a less skilled player. Among amateur tennis players, for instance, the brain circuitry devoted to tennis simply isn't well enough developed to withstand conscious attention directed during competitive play to the separate components of their tennis skills.

But be forewarned: Such gimmicks usually work only against amateur players. They're ineffective against professionals, because after thousands of hours of practice the brain of the true professional has established and honed the relevant circuits until nothing interferes with their autonomous operation.

Puzzles provide an excellent means of enhancing perceptual learning skills and, in the process, activating different brain areas, depending on the chosen puzzle. Here is a puzzle that activates visual-spatial thinking:

PERCEPTUAL SHIFTS *(Visual Thinking)*

ANSWERS ON PAGES 94–95

▶ *This puzzle strengthens your ability to scan for patterns and flexibly changes the way you perceive a figure.*

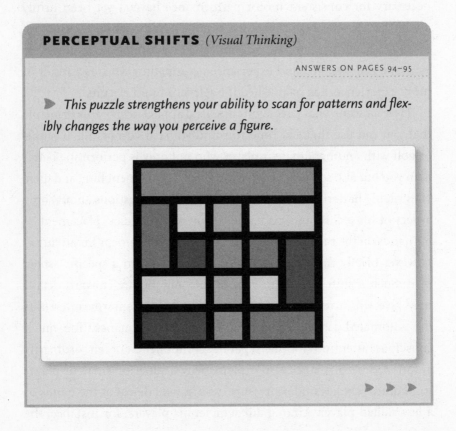

▶ ▶ ▶

I devised this logo for my puzzle-designer friend Terry Stickels. He uses it on his stationery and website. Terry has written many books, and his puzzles appear in *USA Today*'s *USA Weekend* Sunday newspaper magazine. Terry particularly enjoys designing puzzles that involve geometric thinking and shifts in perception, so I designed him a logo that does both. Notice how your perception of this logo keeps shifting as you work through the following puzzles.

① What two letters are superimposed in this logo?

② How many tiles make up this mosaic? Some tiles are squares; some are rectangles. Do not count square or rectangular shapes that are made up of smaller tiles.

③ If the black lines are roads, how many T intersections are there in this map? Hint: How can you count the T intersections systematically so you make sure you don't miss any and don't count an intersection more than once?

④ How many squares of any size are contained in this design? Some squares will be individual tiles; some will be made of more than one tile. Hint: I recommend you first count the smallest-size squares, then count squares that are twice that size, three times that size, etc.

⑤ How many exact copies of the T shape below appear in the design? The figure may appear in any rotation.

REFLECTION. How did it feel to shift your perception from one puzzle to the next? Can you think of other things to count in this figure?

ANSWERS TO PERCEPTUAL SHIFTS

❶ The two letters T and S, which are Terry Stickels's initials, are superimposed in this figure.

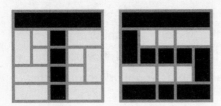

❷ There are 15 rectangular tiles in this mosaic.

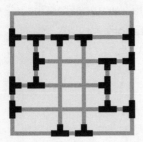

❸ There are 16 different T intersections in this map. A good way to count the intersections is to scan across each horizontal line, left to right, top to bottom.

❹ There are 17 squares in the figure below. I've listed the squares in order reading left to right, top to bottom, which is the best way to make sure you catch every square.

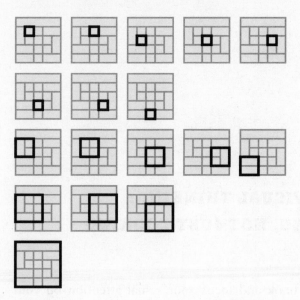

❺ There are 7 different T shapes in this figure.

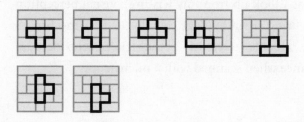

4

VISUAL THINKING:
SEEING, NOT JUST LOOKING

LOOK UP from this book and focus your visual attention on your surroundings. No matter how simple the scene (perhaps limited to the confines of your living room or den) or complex (like the woods I see across the street as I look up from my writing), visual perception originates from the firing of individual cells located in the brain's primary visual cortex, an area referred to as the striate cortex because of its striped appearance when scanned with a microscope.

Seeing Edges: Illusory Contours

When we look at things, the cells in the striate cortex respond to lines of specific orientation, size, and location; edges; and motion. And each of the striate cells is very particular: change a line's orientation by only a few degrees one way or the other and the cell will cease to respond.

Since we don't see the world in terms of lines, edges, orientation, and motion, our brains must bring additional parts of the visual cortex into play. Activation of the cells immediately surrounding the striate cortex leads to increasing detail but also leaves us susceptible to visual illusions such as that illustrated in the figure bélow, which is an example of a Kanizsa figure, named after its creator, Gaetano Kanizsa (1913–1993), an Italian psychologist who had a lifelong interest in illusory contours:

While you see both a white and a black triangle, most of the details of the outlines of both triangles aren't really there. In fact, if you really concentrate your gaze on the space between the apex of the white triangle and the base of the black triangle you can make the white triangle totally disappear. Look now at the figure below, which is another Kanizsa figure. Here, the outline of a house can be "seen" against the black background of five black partially completed circles and five straight lines. The missing contours of the circles, along with the five lines that appear to touch the two sloping sides of the roof, trick the brain into perceiving the outline of a house as seen from the side. But no "house" really exists: the figure consists of only the five partially completed black circles and the five straight lines. (Trace it for yourself in order to confirm that there is no physical difference between the white background and the white house.) Thus, "edge neurons" in the striate cortex can't really be detecting the edges of the "house" because there aren't any real edges to detect! These phantom edges, which are perceived even though they're not really there, are called *illusory contours*.

Illusory contours are created based upon the brain's tendency to perceive a line as continuing in what seems to be its apparent direction. Thus, we see a square underneath a triangle in the previous figure, even though its four corners are missing: the brain responds as if envisioning four lines meeting at the square's corners. Visual misperceptions also occur if some part of the figure is hidden or obscured by another figure. In this case the brain fills in the most likely of two or more possible interpretations based on life experience. For instance, simple curves are more frequently encountered by most people than S-shaped curves (think of body parts like the bend of an elbow, or a bend in the road when you drive). These life experiences increase the likelihood that in ambiguous circumstances the brain will "see" a simple curve more quickly than an S-shaped curve. In the figures below, we're more likely to decide that the two straight lines in the figure on the left side must represent two lines, whereas in the figure on the right we can easily imagine a single line in the form of a simple curve with part of the curve obscured by the black rectangle. Think of the brain as a gambler who wagers on the most likely of two or more alternative explanations for an ambiguous figure. As with any gambler, the brain isn't always right, as can be seen in the lower portions of the diagrams, which show a line taking the form of either a smooth or an S-shaped curve.

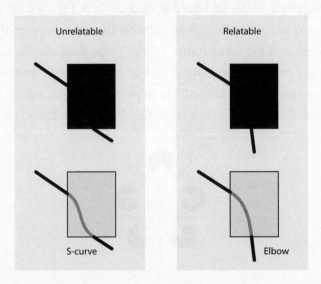

If the brain is susceptible to visual illusions at the level of such elementary perceptions as lines and edges, it should be no surprise to learn that even greater mischief may result from higher-order processing. Attention plays a huge role here. Indeed, normal visual perception isn't even possible without attention. Look at the next diagram below and quickly come up with a description of what you're seeing. Here's my description: Crosses or pluses that combine black and gray in various vertical and horizontal combinations. But wait—a deeper description is possible. In this array, there are two and only two configurations in which both of the verticals are black. See how long it takes for you to find them. However long it takes, you will have to pay attention and look carefully at each of the forms. This is not a fast involuntary process but a special form of visual processing that demands attention aimed at detecting a rare pattern variation.

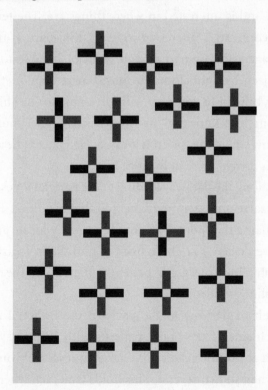

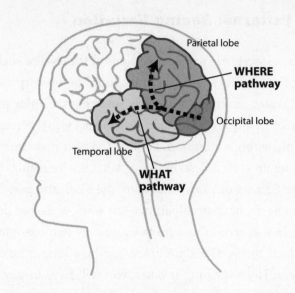

As can be seen from the diagram above, two separate pathways lead from the visual area in the occipital lobe. One passes through the parietal lobe and is interested in *where* things are. The temporal lobe pathway, in contrast, is interested in *what* things are. If the temporal lobe is damaged, the injured person will not be able to name even common objects by sight alone. He may stare at a spoon, tell you what it is used for, but fail to come up with the name for it. This condition, agnosia (from "not knowing" in Greek), spares the injured person's ability to know where the spoon is located. He may not be able to name it, but he sees it *right there* in front of him.

Damage along the pathway leading to the parietal lobes, in contrast, leads to a bizarre and seemingly inexplicable condition in which the patient can name the spoon but cannot correctly locate and reach for it in space, even when it's right in front of him. Neurologists encounter this odd combination of signs in patients afflicted with the rare disorder known as Bálint's syndrome.

As a result of damage to the parts of the occipital and parietal lobes on each side of the brain, patients with Bálint's syndrome are afflicted with an additional disability: they can see only one object at a time, even when shown several. This holds true even when the objects are aligned beside each other. While this fortunately rare condition, dubbed simultagnosia (the inability to perceive more than one visual object at a time), is limited to people with brain damage, under certain circumstances (described below) normal people, too, can experience difficulty visually processing more than one object at a time. That's because two separate processes are at work within the brain.

Seeing Scenes: Change Blindness

Thanks to the first process, we can rapidly scan visual scenes and remember them. Our brain has little difficulty recognizing a picture on seeing it for the second time. For instance, most people, when shown a four-by-four array of sixteen pictures, followed by a second array containing eight repetitions from the first array, will have little difficulty spotting the repetitions. Indeed, some people's performances on such visual puzzles are sometimes nothing less than amazing. In tests involving 612 pictures, volunteer subjects were 98 percent correct, according to the findings of Roger Shepard, who conducted the original picture-recognition test in 1967. Subsequent tests have shown even more impressive performances: 85 percent accuracy can be achieved in response to 2,500 pictures shown for only two seconds apiece. Incidentally, the performance rate for recognizing repeated words from one list to another is far lower. Our brain's enhanced performance with pictures versus words reflects the fact that pictures are more direct and less mediated than words (a picture of a cat versus a description of a cat).

The second brain process requires us to *pay attention* to what we're seeing. To test your ability to do this, go to www.washingtonpost .com/secondglance. This site features a series of photographs matched to altered versions containing a number of changes. Although the pictures seem to be identical on casual inspection, there are actually subtle differences between them. In another version of this challenge, designed by psychologist Ron Rensink, the pictures are shown on a monitor one at a time. The first picture vanishes from the screen for 80 milliseconds (0.08 second) and then is replaced by the second picture. The viewer's task is to detect the changes distinguishing picture one from picture two. The two versions of the same picture then flip back and forth: picture one followed by a blank screen for 80 milliseconds, followed by picture two. The viewers in Rensink's experiment took quite a while to detect the differences, and some people never detected them. Neurophysiologists refer to Rensink's discovery as *change blindness.*

In order to experience change blindness for yourself, check out Ron Rensink's website, www.psych.ubc.ca/~rensink. Scroll down and click on "Visual Cognition Laboratory," then click on "Demonstrations." There you will find ten demonstrations of what Ron Rensink refers to as "the difference between looking and seeing."

At this point it seems that we're encountering a paradox: How can our brain remember hundreds of pictures after seeing them for only a second or two, yet fail to detect obvious differences in only two pictures when looked at over many seconds (and in some cases never detect the changes at all)? Actually, the apparent paradox results from the operation within the brain of two distinctive forms of "seeing." One is rapid and automatic, taking place in milliseconds. The second requires attention and takes longer.

I had the Rensink experiment in mind when I mentioned earlier that under certain circumstances a normal person's response may be very similar to that of a patient afflicted with Bálint's syndrome: the viewer sees only one object at a time. Rensink's experiments suggest a solution to that paradox, as indicated by the title of his original paper: "To See or Not to See: The Need for Attention to Perceive Changes in Visual Scenes." In order to "see" a change, it's necessary to pay attention. And since paying attention takes time, the brain doesn't respond as quickly when searching for subtle differences compared with its rapid response when registering recognition.

In everyday life, we rarely encounter change blindness. But realistic situations can be created by psychologists to demonstrate it. In one of them, psychologist Daniel J. Simons or his collaborator, Daniel T. Levin, asked directions from a passerby on the street. While the passerby was giving the directions, two men walked by, carrying a door. At the instant the door came between the passerby and Simons or Levin, a switch took place. When the psychologist was hidden behind the door, one of the men carrying the door exchanged places with him. In many instances, the passerby failed to notice the change and simply resumed giving directions to the new person. Not only is this experiment an intriguing example of the looking/seeing distinction, it's also amusing to watch. You can view it by going on Google or your favorite search engine and entering the four words *Daniel Simons perceptual blindness*. This will take you, thanks to the Visual Cognition Lab of the University of Illinois, to a video of the experiment.

Visual illusions are fun to play with because they show us that perception isn't always reality. Often, we don't see what is there but rather what *was* there when we were paying attention.

Visual-Thinking Skills

Now, what are visual-thinking skills, why do they matter, and how can you develop your visual-thinking ability?

You may think that visual skill means the ability to draw, but that's only part of the story. As Professor Robert H. McKim explains in his book *Experience in Visual Thinking*, there are three aspects to visual thinking: seeing, drawing, and imagining. The three are intimately connected: When you draw or photograph an object, you come to see it more fully. When you imagine an object, you draw it in your mind's eye. And when you see an object, your eyes scan only a tiny bit of the object at a time; the impression of the full object is assembled in your imagination.

Visual skills matter in fields that deal directly with appearance, such as art, fashion design, illustration, Web design, and filmmaking. Visual skills also matter in fields where much of the communication happens in images, such as architecture, mechanical engineering, molecular biology, and radiology. A new breed of meeting facilitators, called "graphic facilitators," uses improvised sketches to help groups of people work together better.

Basic visual skills, such as the ability to diagram an idea or sketch a plan, are helpful with everyday communication and planning tasks, even if the tasks aren't themselves visual. For instance, many writing teachers have students capture their initial ideas for an essay by drawing a diagram of loosely connected words, rather than writing complete sentences. These "mind maps" (as named by author Tony Buzan) are a mix of visual and verbal thinking.

Styles of Thinking

Before we get to puzzles that develop visual-thinking skills, I want to talk about the interesting split separating people who think in words and people who think in pictures.

We've all met people who are stronger at certain styles of thinking than others. Writers have a gift for words, photographers are adept at images, and accountants have a feel for numbers. In the world of puzzles, most people have a strong preference for one type of puzzle over others, word people prefer crossword puzzles, visual people prefer jigsaw puzzles, and mathematically minded people prefer logic or number puzzles.

People who are good in one area are often weak in others. An artist who paints beautiful pictures may have trouble talking about her work, while a novelist who evokes vivid images through words may have trouble translating his work into a movie script. Of course, some people are strong in several different thinking styles.

Psychologist Howard Gardner popularized the notion of different styles of thinking in his theory of multiple intelligences. His list includes at least nine styles of thinking, any of which can be developed separately.

Visit a bookstore and you will see the split between words and pictures played out in dramatic fashion. In the fiction section, you will find books filled mostly with words. In some of them, a few pictures or illustrations appear to support the text. In the art section, you will find books filled mostly with pictures; words appear only as captions that explain the pictures. Only in the children's section will you find a large number of books with equal emphasis on words and pictures.

But the balance between words and images runs deeper than mere preference. Word people live in a world of language, with all of their experience moderated by an internal dialogue that is difficult if not impossible to silence. Similarly, picture people live in a world of images, with an eye for visual appearance that is difficult if not impossible to turn off. Word people and picture people often find it difficult to believe that the other thinking style exists. But in fact people vary greatly in their preferred mode of thinking.

The split between thinking in words and thinking in pictures isn't black-and-white. Most of us do a little of both. Personally, I find that my mode of thinking is quite fluid, influenced by what I am doing. When I see a movie, I am in my visual mind, aware of lighting, movement, and composition. When I listen to music, I am filled with rhythm and melody. After I read a book, my internal dialogue is particularly strong.

Here are puzzles designed to get you thinking in the visual mode:

HIDDEN DETAILS *(Visual Thinking)*

ANSWERS ON PAGES 114–115

▶ *This puzzle strengthens your ability to scan for visual patterns.*

Here are four picture puzzles. In each puzzle, find where each of the small circular details appears in the big picture. Each detail is right side up and full size.

❶ Harpist

② Blocks

▶ ▶ ▶

③ Snowy river

1   2   3   4   5   6

REFLECTION. The Hidden Details puzzles focus on the two-dimensional pattern of shading and shape in a picture, rather than on naming the objects in a picture. As with solving a jigsaw puzzle, you must look for cues in a small fragment that tell you where it fits in the larger picture.

If you are an artist or visual designer, this exercise will seem familiar. If words are your preferred mode of thinking, then you will probably approach this problem by verbalizing the features of each detail, then searching the picture for those features. The more you do this puzzle, the quieter your internal dialogue will become, and the more directly you will be thinking visually.

The sequence of images goes from easy to hard. Each image presents a different kind of visual challenge.

▶ ▶ ▶

▶ ▶ ▶

4 Graffiti

1   2   3   4   5   6

1 Harpist. This is the easiest image to find details in, because the person and instrument have familiar distinctive parts that are easy to identify and name.

2 Blocks. This puzzle is a bit harder. All the shapes have simple straight edges and solid colors, but they are combined in complex ways.

3 Snowy river. This puzzle is quite difficult. A nature scene like this is full of complex shapes that are confusingly similar to each other, especially in black and white.

4 Graffiti. This puzzle is the hardest of all. The image has a confusingly similar texture everywhere, and there are no identifiable objects or symbols for the eye to hold on to.

▶ ▶ ▶

▶ ▶ ▶

To play more Hidden Details puzzles, go to the website www
.theplayfulbrain.com and play Photograb. This game lets you find
details in photos taken by other people and even make puzzles
out of your own photos to share with friends.

As an exercise in visual thinking, picture in your mind for just a
moment a mental image of your house, condominium, or apartment
building as seen from across the street. Make the image as clear and
detailed as you can. If the mental image is unclear, walk across the street
and take a careful look at the building in order to gather additional
details for your mental image. Next, look at a chair in your living room
and then turn away and construct a mental image of it. Finally, stare
at your face in the mirror for a moment and then close your eyes and
re-create that mirror image.

In each of these exercises, specific regions are activated in the
ventral (lower) temporal cortex according to the category of objects
involved: houses, furniture, and faces. No one knows why the brain is
organized in this category-specific fashion. At one time neuroscientists
spoke of a "grandmother cell," a jocular reference to the theory that
the brain contained cells that activated only when you encountered
your grandmother. Today that theory is considered naïve: How could
the activation of a single neuron possibly represent the aroma of
your grandmother's cookies along with all of the other wonderful
features associated with your experience and/or memory of your
grandmother? It's much more likely that the brain's representation of
your grandmother and other specific persons and objects in your life
involves the activation of many functionally interlinked brain regions,
or *modules*.

But whatever the explanation for category representation, it is
sufficiently precise that a neuroradiologist can predict by scanning your
fMRI whether you are looking at a building, an item of furniture, or

a face. Even more interesting, that neuroradiologist can make equally accurate predictions when you are simply *picturing in your mind* the building, chair, or face. That's because those same activation patterns occur within the brain when you're not actually looking at pictures of houses or faces but merely *imagining* them. In other words, the same regions involved in the perception of an object are also involved in the formation of mental pictures of the object!

This finding, which has been confirmed by several fMRI studies, points to two important points about the brain.

First, images are stored in the same regions of the brain that are activated during the initial visual processing. Second, remembering those images activates the same brain areas that were active during the original experience (in this case, looking at the building, the furniture, or the face). Such discoveries validate many of our contemporary ideas about skill learning. Although practice is important—you can learn a skill only by actually doing it—mental imagery can also play an important role. Athletes in all major sports now study videos of their performance and, on the basis of perceived deficiencies, create vivid images of how they should have responded. These mental "snapshots" are then used to formulate corrective performance changes.

This finding—that similar brain areas are involved in perception and imagination—opens up a whole world of brain and memory enhancement. Imagining a successful outcome of an activity activates the same brain circuitry involved in successfully completing that activity. Now, that isn't the same thing as claiming that by mentally envisioning yourself winning your next game against your toughest tennis opponent, you will actually be able to do so. But this work on mental imagery and its correlation with specific brain areas suggests a role in brain enhancement for certain puzzle challenges involving mental rotation.

For one thing, the reaction time differences required for solving puzzles supports the temporal activation theory of how the brain works. In general, the more complex the mental processing required to solve the puzzle, the more time required to come up with the solution. This correlation of time to difficulty is unlikely to result from the simple linkage of more and more neurons or neuronal circuits. Thanks to the speed of information transfer within the brain, it shouldn't take that much more time to solve a difficult puzzle than an easy one if nothing more is involved than simply recruiting more neurons or more neuronal circuits. It's more likely that difficult puzzles require more time to solve because they require synchronization of increasingly widespread neuronal circuits.

How vividly can you imagine an apple if you close your eyes and try to picture it? If you are a strong visual thinker, you will imagine the apple in great detail, down to its particular shape and skin blemishes. People who work with images all day, such as photographers, have no trouble recalling particular images in great detail, even manipulating and combining images in their imagination.

As mentioned earlier, we now know from neural imaging that when you imagine seeing an object, your brain activates the same areas of your brain as if you were actually seeing it. (That goes for other senses as well.) It's as if you were running your vision system in reverse: from the concept of an object, your brain constructs the electrical impulses that would have fired had you actually seen it.

Here is a puzzle that will strengthen your ability to recall the appearance of some everyday objects:

ANSWERS ON PAGE 115

▶ *This puzzle strengthens your ability to notice and recall visual detail.*

Draw the following items from memory. Then check each drawing against the real thing or a picture of the item and try the exercise again. The more you practice drawing from memory, the more detail you will recall.

1. **Stop sign.** How many sides does the sign have?
2. **American flag.** How many stars and stripes? How are they positioned?
3. **Apple Inc. logo.** What details of the apple shape are included? Is there a stem? A leaf? If so, which way do they face?
4. **The capital G in the General Mills logo.**
5. **The complete capital cursive alphabet you learned as a child.** Which letters are hardest to remember? Note: People raised in different countries learn slightly different letter shapes.

REFLECTION. Which figures were easy to recall? Which were hard? Which details were hard to remember? Ask friends to try this puzzle and see what details they remember.

Y

1

2

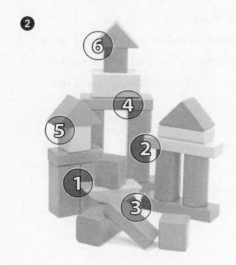

③

④

ANSWERS TO DRAWING FROM MEMORY

1.

2.

3.

4.

R

5

SPATIAL THINKING:
THE CHALLENGE OF MENTAL ROTATION

SPATIAL THINKING is closely related to visual thinking. But whereas visual thinking deals with flat, two-dimensional images, spatial thinking deals with three-dimensional shapes—the world we inhabit as living beings moving through a three-dimensional world. For instance, when you drew the stop sign from memory in the last chapter, you didn't have to think in three dimensions or integrate the mental picture of the stop sign as you remembered it in your mind's eye with how that stop sign would feel if you touched it.

Spatial thinking involves a peculiar paradox. We are born into a three-dimensional existence. Children's games include clambering around fully three-dimensional climbing structures and building with three-dimensional construction toys like Legos. But our schools neglect visual and spatial thinking, focusing instead on developing language and mathematics skills. As a result, far more people think in words or two-dimensional images and are not very good at spatial thinking. This isn't universal, of course: Spatial thinking in three dimensions is a prerequisite in fields like carpentry, sculpture, and architecture. Surgeons and molecular biologists also need highly developed spatial-thinking skills, since they must be able to understand and manipulate bodily organs in three dimensions. But the greater majority of the population rarely deals with fully three-dimensional problems that require spatial visualization and manipulation. Because of this imbalance, gifted visual-spatial thinkers are often seen as less intelligent than they are. To make up for this imbalance, we will give extra attention to spatial-thinking puzzles in this chapter.

To experience spatial thinking, pick up a small but interesting object such as a ring. Study it carefully. Move it around in your hand. Imagine what the ring would look like from the side or upside down. Imagine yourself making the hand and finger movements necessary to bring about those changes. Close your eyes and picture the ring vividly. Then, with your eyes still closed, rotate the ring in your hand and try to visualize in detail how the object now looks. Open your eyes and see if your visualization is correct. What you have just done, especially with the three-dimensional rotation, is an exercise in visual thinking.

Before presenting puzzles aimed at developing spatial thinking, we'll explore what's going on in the brain when you perform mental rotation puzzles.

Thanks to recent breakthroughs in neuroimaging via fMRI, we now know that mental rotation activates a wide swath of brain tissue (much of it in the right hemisphere) that isn't used in reading, speaking, writing, and other predominantly left-hemisphere activities. Included here are areas governing eye movements and vision, along with the control areas in the parietal and motor areas responsible for sensation and movement of the hands. One of these motor areas, the supplementary motor area, or SMA, is activated in any difficult task involving substantial attention, such as mental rotation.

According to the neuroscientists who carried out the fMRI testing, the machinery of primary sensation, imagery, and perhaps perception might well be identical. This suggests yet another reason for working mental rotation puzzles: All three of the mental faculties mentioned can be improved simultaneously.

Here are puzzles that will help you develop your spatial-thinking ability. Let's start with the most basic building blocks of spatial thinking: squares and cubes. After that, we'll move on to more complex shapes and operations.

MENTAL BLOCKS *(Spatial Thinking)*

ANSWERS ON PAGES 139–141

▶ *This puzzle strengthens your ability to manipulate shapes in your imagination.*

❶ Which shapes below can you make by assembling the two pieces at right? You may move the pieces around, but you may not turn or overlap them. To get you started, I've shown you how to assemble shape 7.

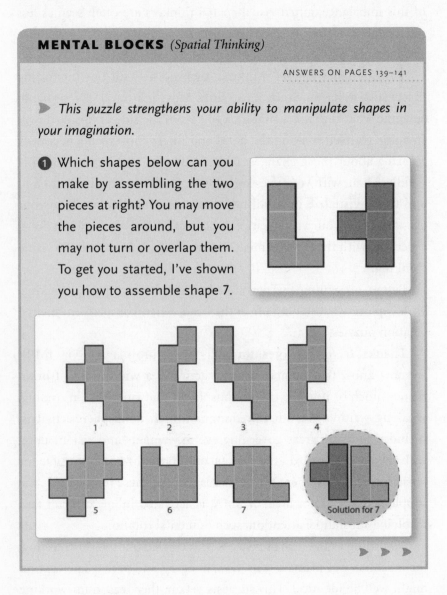

▶ ▶ ▶

❷ Which shapes below can you make by assembling the two pieces at right? You may turn or flip pieces, but not overlap them. To get you started, I've shown you how to assemble shape 7.

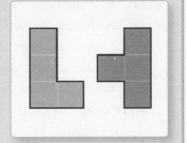

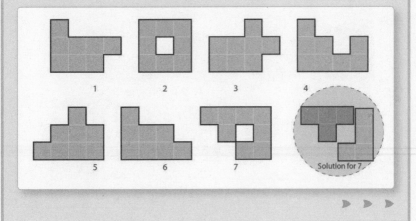

1

2

3

4

5

6

7

Solution for 7

▶ ▶ ▶

P

▶ ▶ ▶

3 Which shapes below can you make by assembling the two pieces at right? You may rotate the pieces in three dimensions. Some cubes may be hidden from view. To get you started, I've shown you how to assemble shape 7.

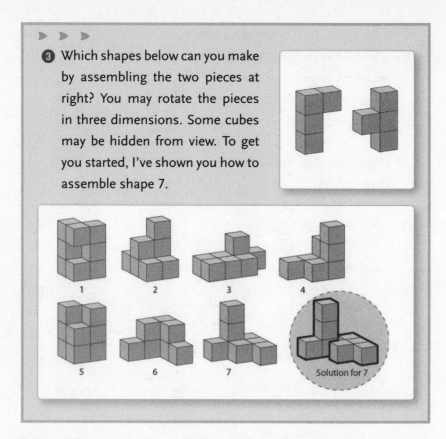

RICHARD What has been your experience, Scott, in designing your mental rotation puzzles? Why do some people find them reasonably easy to do, while others, including me, find them almost impossible?

SCOTT Over many years of designing puzzles, I've learned that those requiring the visualization of shapes are sharply divisive: some people love them, and some people hate them, mostly because they feel they have no ability in that area.

I believe that spatial visualization skills—like all cognitive skills—are quite learnable. Certainly, some people have more natural ability than others. But everyone can improve their spatial visualization ability. Here's how.

First, get your hands dirty: Build things. Spatial visualization is as much a motor skill as it is a perceptual skill. Working with building blocks like Legos, cubical blocks, or other building blocks is useful because their simple proportions help your brain grasp spatial relationships. When I was a child I spent hours building with wooden blocks. I constantly asked myself questions like *How many of this block equals that block?* and *What will I get if I turn that shape on its side?*

Next, draw pictures of 3-D structures so that you understand the relation between flat drawings and 3-D objects. Architects and engineers learn these skills in school using pencil sketches and computer-aided design tools. Artists, too, learn a great deal about space when they do life drawing. In particular, they learn to simplify complex 3-D forms into basic shapes like spheres, cones, and cylinders.

Finally, develop a vocabulary for talking about 3-D shapes and transformations. The basic forms include the cube, sphere, cylinder, cone, and prism. The basic spatial transformations include translation (moving an object in a direction without turning it), rotation (turning an object), reflection (flipping an object in a mirror), and scaling (changing the size of an object). Then there are operations for changing one shape into another, such as extruding a cross section to make a long straight shape, and revolving a cross section around an axis to make a cylindrical shape. Be sure to connect these words with experiences of building and drawing things: students who learn these ideas in school as purely abstract ideas usually fail to master them.

RICHARD Your advice is especially important for people with underdeveloped skills in spatial processing who rely too much on words and concepts. Too much verbal thinking interferes with mental rotation exercises. This interference is especially true for people like writers who think primarily in words. The same can be said for neurologists like myself as well as just about every other non-surgeon. My mental rotation skills aren't that highly developed. Indeed, my limited mental rotation ability is one of the reasons I chose to become a neurologist rather than a neurosurgeon. As a neurologist, I can consult a textbook in my office when I need illustrations of the relevant brain anatomy germane to a particular patient's problem. If I had taken up neurosurgery, I would be forced to visualize everything while standing over my patient in the operating room.

SCOTT A verbal approach will get you a certain distance with mental rotation, but you'll do better if you think directly in visual terms. You have to take deliberate steps to make visual language something you think in directly. It's like learning a foreign language: At first, you continue to think in your native language and translate words one at a time into the other language. As you become more fluent, you learn to think directly in the new language, without going through the extra translation step.

RICHARD When I took up the challenge of increasing my mental rotation skills, I looked for some assistance. I've found photocopiers and mirrors very helpful in improving my mental rotation skills, especially when approaching harder puzzles that require precise mental visualization.

~

When you've become comfortable with these warm-up exercises here's a more challenging puzzle called Pentominoes. It's a valuable exercise that is well worth the investment of time.

ANSWER ON PAGE 141

▶ *This puzzle strengthens your ability to manipulate shapes in your imagination.*

Here's a puzzle that will improve your mental rotation abilities as well as other aspects of your spatial visualization skills. The puzzle is challenging but well within most people's abilities. Stick with it and your effort will be rewarded. It can take up to a couple of hours to solve; many people find they need to put it aside and keep coming back to it a little bit at a time over the course of a day.

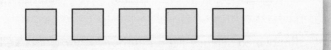

You will need five identical squares, up to six inches on a side. Use whatever is convenient—dice, coasters, Post-it notes—or cut your own out of cardboard or paper. You will also need a pencil and paper for recording your solution.

Join the five squares along edges to make a single shape. Adjacent squares must join along an entire edge. Squares may not overlap or touch only at corners. For instance, here are two different shapes you could make:

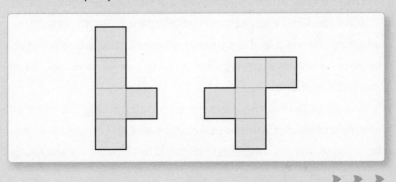

▶ ▶ ▶

L

▶ ▶ ▶

Shapes made of five squares are called "pentominoes," pronounced to rhyme with *dominoes*. *Penta* means "five," as in *pentagon*.

Your challenge is to discover all the different pentominoes. Shapes that differ only by rotation or reflection are not considered different. For instance, the following three figures all count as the same shape.

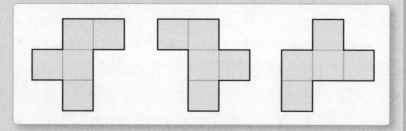

Renowned science fiction author Arthur C. Clarke liked this problem so much that he devoted a chapter to it in his novel *Imperial Earth*. The protagonist works through the puzzle over the course of a day and keeps discovering new shapes he has missed.

Tips. It's easy to discover several shapes but hard to know whether you have found them all. You may want to put this puzzle aside and come back to it later, or ask a friend to see if he can come up with any other shapes.

As you work on this puzzle, you will be exercising your spatial visualization skills, especially mental rotation, as you determine whether shapes are the same or different. You will also need to come up with a strategy for figuring out whether you have missed any shapes.

As an easier warm-up exercise, try determining the number of different shapes you can make with just four squares. Incidentally, these shapes are called "tetrominoes" (*tetra* means four).

▶ ▶ ▶

▶ ▶ ▶

These shapes are similar to the shapes in the popular computer game Tetris. It is no coincidence that the name is similar to the prefix *tetra*: the inventor of Tetris, Alexey Pajitnov, is a mathematician who was inspired by pentominoes.

HINTS FOR PENTOMINOES

Here is a strategy for making sure you have found all the pentominoes. The general problem-solving strategy is to break a complex problem into simpler problems that can each be solved separately.

The strategy I use to solve this puzzle is to order the pentominoes by the longest sequence of consecutive squares in a straight line within the figure. There is one pentomino with five squares in a line, then there are two pentominoes (not hard to find) with four squares in a line. In fact, a good way to find those two pentominoes is to draw (or make) four squares in a line, then consider all the places where an extra square can be added on. There are ten such places. Two create the five-in-a-row pentomino, and the remaining eight collapse to two when you ignore shapes that are symmetrical copies of each other.

That leaves three-in-a-row pentominoes. Enumerating these is considerably harder, and there are a few that are easy to miss; remember that a pentomino might have two different three-in-a-row sequences.

Finally, there are pentominoes that contain only two-in-a-row sequences. Can you find how many there are? This one is not too hard to solve.

A fun way to explore 3-D visualization is to make shapes using your arms and hands. Here are some shape games, excerpted from *Math Dance with Dr. Schaffer and Mr. Stern.*

THREE-PERSON HANDSHAKE
(Spatial Thinking, Creativity)

(Adapted from Math Dance with Dr. Schaffer and Mr. Stern *by Schaffer, Stern, and Kim)*

▶ *This game challenges your ability to notice and create symmetrical patterns.*

Gather three people. Face each other in a circle. Invent a handshake for all of you to do. The only rule: All of you must do exactly the same thing. For instance, if one person crosses her right arm over her left, the others must do the same

Here are a couple of the many solutions:

Photographs by Steve Savage

REFLECTION. Is your figure exactly symmetrical? How did you check? Did each person use one hand, both hands, or no hands? Can you invent two more three-person handshakes that are different from the first? Can you perform the three handshakes one after another without stopping or talking to each other?

MIRRORED POSTURE *(Spatial Thinking)*

(Adapted from Math Dance with Dr. Schaffer and Mr. Stern *by Schaffer, Stern, and Kim)*

 This puzzle challenges your ability to notice and create symmetrical patterns.

Here's a quick, fun exercise that will make you aware of the position of your body in space and thereby sharpen your sense of body image.

Whatever position you are in at this moment, freeze! Now notice where all the parts of your body are. How are your hands and feet positioned? Is your head tilted to one side? Is your weight shifted one way or the other? If you are holding this book, how are your hands arranged?

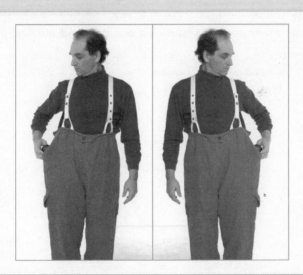

Photograph by Steve Savage

▶ ▶ ▶

Mentally record your body position, then reposition your body with left and right reversed. If you were holding this book with your left hand, hold it with your right. If you were leaning to the right, lean to your left. You may find that the new position feels odd: most people prefer to do certain things, such as crossing their legs, one way rather than the mirror-image way.

REFLECTION. What did you notice about your body position? Did it feel odd to hold your body in the mirrored position?

Throughout the day, try reversing left and right. Hold your spoon with your opposite hand as you eat. Reverse the foot you put a shoe on first. Fold your arms the "wrong way" (this can be challenging).

You can also play this game with a friend. One person assumes a position; the other mirrors it as quickly as possible. Then the other person assumes a position and the first person mirrors it.

TWO-PERSON TETRAHEDRON *(Spatial Thinking)*

(Adapted from Math Dance with Dr. Schaffer and Mr. Stern *by Schaffer, Stern, and Kim)*

▶ *This activity exercises your ability to understand and create three-dimensional shapes.*

The simplest three-dimensional shape is the tetrahedron, which looks like a pyramid. Yes, the cube is better known, but if you want to understand three-dimensional structures, the tetrahedron is just as important.

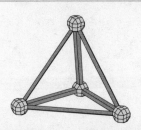

The tetrahedron is a pyramid, but unlike the great pyramids in Egypt, which have square bases, this pyramid has a triangular base. Notice that the tetrahedron has four triangular faces—one on the bottom and three around the sides—and six edges—three around the bottom and three meeting at the top.

▶ ▶ ▶

Here is a fun way for you and a friend to make a tetrahedron using only your fingers:

Face each other. Each of you, join your thumbs together. Straighten out your thumbs so together they make one straight line. Then extend your index and middle fingers so they point away from you, and the four fingertips touch the corners of an imaginary square.

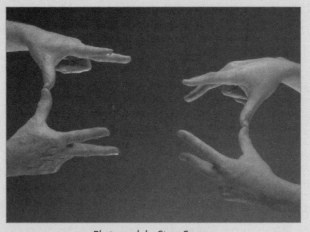

Photograph by Steve Savage

▶ ▶ ▶

▶ ▶ ▶

Now one of you, but not both: Twist your hands sideways, keeping your thumbs together, so one hand is directly above the other, instead of being side by side.

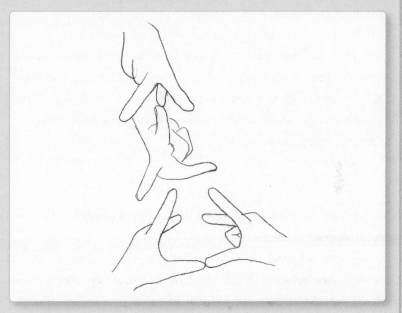

Now bring your hands together and join fingertips. You will get the following shape.

Photograph by Steve Savage

▶ ▶ ▶

A

▶ ▶ ▶

Straighten out your fingers so each pair of fingers that meets makes a straight line. Look carefully and you will see that there are six lines, each made of two fingers, and four triangles, each made of six fingers: a tetrahedron!

REFLECTION. How easily can you see the tetrahedron shape? Can you see the six lines? The four corners? The four triangles? What other shapes can you make with your hands? With three people you can make a larger tetrahedron, using your forearms. Can you figure out how?

Now that we have explored three dimensional shapes, let's turn to three-dimensional rotation.

According to work done decades ago by pioneering Stanford University psychologist Roger Shepard, the more an object must be mentally rotated to determine if it is identical to a second object, the more time will be required to carry out that mental rotation. Put another way, there is a direct correlation between the time required to take the object in one's hands and directly turn it, and the time required for mentally performing the same maneuver. Further, it doesn't matter on which axis the object is rotated; rather, what matters is the extent of the rotation mentally carried out.

In two dimensions, an object can rotate either clockwise or counterclockwise. In three dimensions, rotation is more complex: not only can an object rotate clockwise or counterclockwise, it can also rotate around several different axes. For instance, an airplane can rotate three different ways, called pitch (i.e., it can raise or lower its nose), yaw (it can point its nose to the left or right), and roll (it can twist while keeping its nose pointed forward).

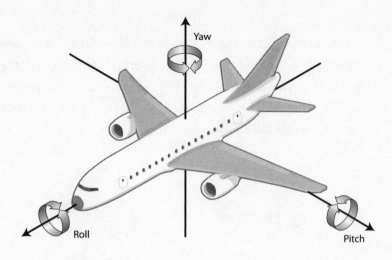

Similarly, dancers and tumblers have names for the three ways they can rotate their whole body: somersault, pirouette, and cartwheel.

Here is a puzzle that will strengthen your ability to imagine three-dimensional rotation.

TURN OF THE CENTURY *(Spatial Thinking)*

ANSWERS ON PAGE 142

> *This puzzle strengthens your ability to visualize three-dimensional rotations.*

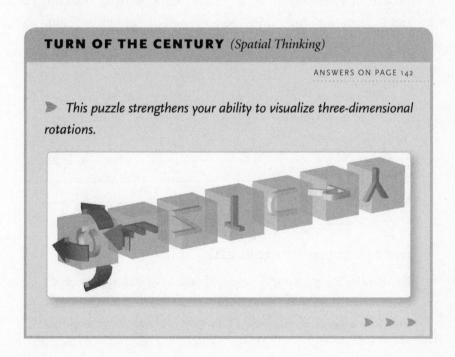

▶ ▶ ▶

F

▶ ▶ ▶

In the picture on the previous page, the seven letters of the word CENTURY are suspended in cubes at different angles. Selecting one of the four arrows surrounding a cube turns the letter 90 degrees left, right, up, or down. For instance, the example below shows the effect of turning right, then up.

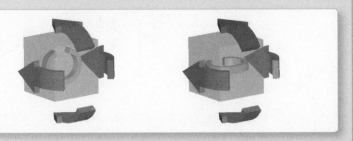

Your challenge is to turn every letter back to its right reading orientation in the fewest possible moves. For instance, C takes two moves: up, then left (or down, then right). Write your answers in the blanks below. Write L for left, R for right, U for up, and D for down. Then write the total number of moves. Hint: No letter takes more than four moves; some have more than one solution.

C _____

E _____

N _____

T _____

U _____

R _____

Y _____

TIPS FOR SOLVING THIS PUZZLE

Try manipulating physical props. If you have 3-D letters, such as children's toys, hold one in your hand and turn it to help you visualize what happens when you turn a letter. To strengthen

▶ ▶ ▶

▶ ▶ ▶

your 3-D visualization skills, try to visualize the result of each turn before you actually turn the letter.

If you don't have physical letters at hand, cut letters out of cardboard. Or write a letter on an index card and hold it in your hand. Or you can pretend to hold a letter in your hand and turn it until it is in the right orientation.

The finished shape of a product like a bowl or a chair is influenced not just by what it does but also by the tools used to make it. For instance, vases made of clay are usually round shapes that look circular when seen from the top. That is because they are formed on a potter's wheel, which spins in a circle as the potter forms the clay.

Photograph by Aloha Lavina

Here is a puzzle that asks you to visualize cylindrical shapes formed by this sort of spinning motion.

REVOLUTIONARY THINKING *(Spatial Thinking)*

ANSWERS ON PAGE 143

▶ *This puzzle strengthens your ability to visualize three-dimensional rotations.*

Notice any unusual similarity between the bottle and the plate in this scene? Both are cylindrical shapes that can be formed on a potter's wheel or lathe. Such cylindrical shapes are called "surfaces of revolution."

But there is more. If you cut open the two shapes, you will see that both were created by spinning cross sections of exactly the same shape (solid line) around an axis (dotted line). For instance, the tapered neck of the bottle and the sloping lip of the plate were swept out by the same part of the cross section. The only difference between the two shapes is that the cross section has been rotated 90 degrees relative to the axis of rotation.

▶ ▶ ▶

▶ ▶ ▶

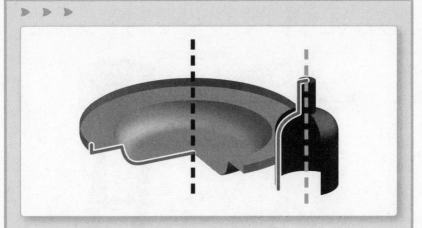

Most 3-D computer modeling programs used in 3-D computer animation and video games include tools for making surfaces of revolution. People who work with these tools become adept at imagining what shape will result if they spin a particular cross-sectional curve around a particular axis.

On the next page are twelve more surfaces of revolution. Six pairs have identical cross sections. Each cross section is made of just two or three straight lines. For each shape, see if you can write in the name of the other shape with the same cross section. Write down your answers. The answer to shape f is shape j, and the cross section is a simple L.

Hint: To find the cross-sectional shape of an object, first find the axis of revolution. Usually the axis is vertical, but sometimes it is horizontal. Then imagine taking a knife and cutting a pie slice out of the object, with the tip of the knife traveling along the axis line. You will be left with three-quarters of the object intact, as shown in the diagram above. Imagine the shape of the edge exposed by the knife: there will be two copies of this edge, one on either side of the pie slice. That edge is the cross-sectional shape you are looking for.

▶ ▶ ▶

N

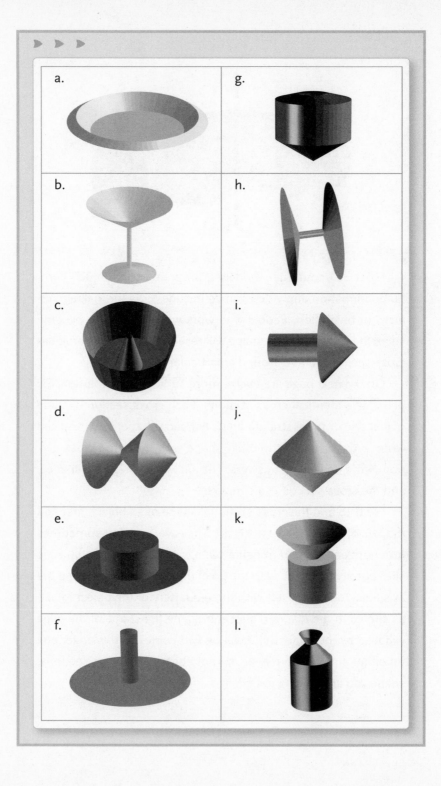

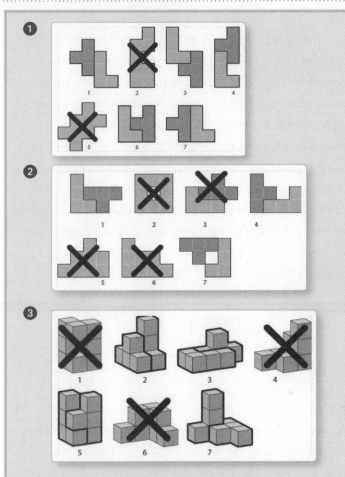

REFLECTION. You probably found the second and third mental blocks puzzles harder than the first. That is because in the second and third puzzles you have to imagine rotating pieces, first in two dimensions, then in three. Rotating shapes in your imagination calls upon additional brain circuits. The more rotations you have to work with, the greater the challenge to your brain.

In order to solve the third puzzle, you can proceed in several ways. First, you can construct a model by cutting out the two figures in the upper right of page 120 (or, better yet, drawing

▶ ▶ ▶

B

▶ ▶ ▶

and cutting a larger version of the figures) and then shifting them around on a table in order to decide on the similarity of the shapes. Second, you can perform the entire operation in your imagination. This will require the following steps: creating a mental image of the patterns; mentally rotating the images until you can compare them; and finally, deciding which of the six shapes can be made by rotating the two pieces. Since this is hard for language-based thinkers, here are several approaches to aid skill enhancement in mentally maneuvering the shapes.

A photocopier or a mirror is especially useful in puzzles like these in which shapes that vary only in rotation or reflection aren't considered different. For instance, the following three figures all count as the same shape:

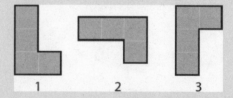

The photocopied version (2) represents the shape of the figure 1 after it's been lifted from the paper and flipped over on its opposite side. Version 3 represents the arrangement when placed before a mirror. Since both the photocopied and the mirror versions represent the identical arrangement but seen from different perspectives (rotation and reflection), they don't count as separate shapes.

After several hours of working with models, mirrors, or photocopiers, even heavily language-dependent thinkers will improve to the point that such puzzles can be solved by mental rotation alone.

▶ ▶ ▶

► ► ►

The more you manipulate shapes with your hands, the easier it will become to imagine these operations in your mind. When I visualize three-dimensional forms, I don't just picture what they look like: I also imagine how they feel. I sense what it would be like to hold them in my hands, feel their weight, run my fingers around their edges, and turn them around.

ANSWER TO PENTOMINOES

Here are the twelve possible shapes that can be made out of five squares.

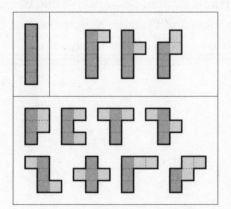

In summary, I used a divide-and-conquer strategy to break the pentomino problem down into simpler problems. The three-in-a-row problem is pretty hard, but the others are pretty easy.

This strategy may seem a little messy, but it's the best one can do. There is no neat mathematical formula for the number of shapes with n squares: you just have to work it out.

REFLECTION. What strategy did you use to solve this puzzle? Can you use the same strategy to count the number of tetrominoes (figures made of four squares)?

Here are all the solutions for each letter. For instance, there are four different three-move sequences for R.

    C: UL, DR
    E: RU, RD
    N: RUR, RDR, LUL, LDL, ULU, URU, DLD, DRD
    T: RUU, RDD, LUU, LDD, UUL, UUR, DDL, DDR
    U: LUL, LUR, RDL, RDR, ULU, DLU, URD, DRD
    R: URR, ULL, LLD, RRD
    Y: UU, DD

REFLECTION. How hard was this puzzle for you? Many people find this puzzle to be quite hard. What techniques did you use to keep track of the orientation of each shape as you turned it in your imagination?

Here are the pairs of shapes that match. I've cut open each shape so you can see its cross section. The dotted lines give the axis of revolution for each figure.

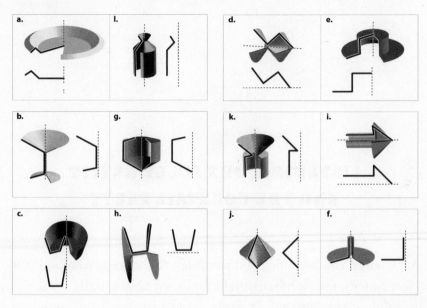

REFLECTION. What techniques did you use to solve this puzzle? How hard was it for you? Most people find this puzzle very hard. Look around your house and find objects that are surfaces of revolution, and trace the shape of the cross section by running a finger along the surface of the object.

6

LISTENING: THE FOLEY ARTIST
AND THE COCKTAIL PARTY

STOP WHAT YOU are doing and just listen to the sounds around you. Near sounds, far sounds, natural sounds, artificial sounds . . . Sound is omnipresent. We can't as easily shield ourselves from sounds as from sights: We don't have "ear lids." Nor do we have to be looking at something to hear it. Thanks to the individual shape of our ear's pinnae, certain sound frequencies are funneled into our ears better than others. Even when we are asleep, our ears are listening.

Sound conveys place. Close your eyes. The sound of the environment will tell you whether you are outside or inside, the size of the space, and the sound-reflective qualities of nearby surfaces. Of course, outdoor environments have sounds like birds and wind rustling the leaves on nearby trees. But even an empty room has sounds. Ambient sound is so important that when filmmakers shoot a movie on location, the sound recording engineer always records the "room tone": the sound of the environment absent any intentional sounds. Later, when the sound effects for the scene are being mixed with the recorded dialogue, the room tone can be added back in. Without room tone, the dialogue and accompanying sounds seem fake; they don't seem to belong to the environment depicted on the screen.

Although vision is our primary sense, as judged from the greater expanse of brain devoted to conveying information from the eyes, sound is the primary conveyer of emotion. Talk to someone who has "witnessed" a horrifying event: If he was close enough to hear anything, it's the sounds he will describe to you. For instance, after witnessing her first street fight, a patient spoke of how upset she was at the sound of teeth breaking and bones shattering. The same thing holds true in storytelling. Try watching a horror movie on television with the sound off but the picture on. Pretty silly, you'll conclude. In movies and on television, sound is also the primary conveyer of meaning. Watch a sitcom with the sound on but the picture off. Chances are that you can follow the storyline just fine: most TV shows rely primarily on sound to carry the story. The reverse is not true: Turn off the sound, and a television program becomes much harder to make sense of.

We usually take sounds for granted, focusing on the sounds that are most important to us at the time while ignoring the other sounds that are around us. This tendency is augmented by what acoustic scientists refer to as continuity and restoration effects. If you use a cell phone, you encounter these effects every day. Think back to the last time you were on a cell phone, getting directions from a friend, and traffic noise or a loud nearby conversation momentarily blocked your hearing of a few of your friend's words. Unless you weren't paying attention in the first place or the interruption lasted too long, you were able to "hear through" the interruption. Your auditory system and brain were able to reconstruct that portion of your friend's speech that was blocked out by the interrupting sound.

Remarkably, as you've probably noticed, the same thing doesn't happen on those occasions when your cell phone's reception momentarily falters as you enter a "dead zone." Even though you may rapidly move back into the reception zone and reestablish your connection with your friend, you'll likely need to ask him to repeat the last sentence. That's because when a gap in the free flow of spoken words is interrupted, silence is harder to "hear through" than noise. You can experience this continuity-and-restoration-effect paradox for yourself by going to www.sinauer.com/wolfe2e, the companion website for Jeremy M. Wolfe's marvelously instructive and entertaining book *Sensation & Perception.* Click on chapter 10, "Hearing in the Environment," and select "10.5: Continuity and Restoration Effects."

Even though this perceptual restoration effect for noise but not silence can be very useful (as with the first cell phone example mentioned above), it has a downside: it makes us susceptible to hearing things that haven't actually been said.

In an experiment originally conducted in 1971 and cited by Wolfe in his textbook, volunteers listened to the sentence "The state governors met with their respective legi*latures convening in the capital city," with the first *s* in "legislatures" replaced by silence, a cough, or noise (represented by the asterisk). Despite the elimination of that *s*, none of the listeners noticed its absence. Even when forewarned that something would be missing from the sentence and replaced with either silence or another sound, none of the volunteer subjects was able to identify where the sentence had been changed, except when silence replaced the missing *s*.

The words we hear can also be determined for us by information we get after we hear those words. We know this on the basis of an experiment by the same scientist who carried out the missing *s* experiment described in the last paragraph. The incomplete words in "The *eel fell off the car" and "The *eal fell off the table" are heard as "wheel" in the first sentence but as "meal" in the second sentence, even though the necessary context for each sentence isn't provided until after the missing phonemes. The brain processes the sentences within the language areas in parts of the temporal and frontal lobes. Simultaneously it holds the missing information gap in working memory, mediated principally by the frontal lobes, until the context of the sentence becomes clear: wheels fall off cars, while meals fall off tables. Most intriguing, all of this takes place outside of conscious awareness.

Of all the sounds surrounding us, language is the most powerful motivator. Compliments, insults, and shouts for help elicit distinct responses. Whatever our native language, it places certain restraints on us that make learning a new language more or less difficult depending on our age when first starting. Infants and young children learn quickly because of their brains' plasticity. Adults, in contrast, have to overcome a lifetime of acquired acoustic sensitivities and language habits that interfere with learning a new language. For example, Japanese adults find it difficult to distinguish the sounds of *r* and *l* because these sounds are both very similar to one sound (called a "flap") in Japanese. Since they never have to distinguish the sounds of *r* and *l* in their native Japanese, their brains fail to register the distinction. As a result, when as an adult the native Japanese speaker tries to learn English, where the distinction is important, he has a hard time enunciating the two sounds. Japanese children do much better at learning English because of their brains' plasticity: the tendency to ignore the *r/l* differences hasn't as yet been firmly established in the form of neuronal circuits.

Below are a few puzzles that illustrate different aspects of how our brain processes sound. The first is aimed at heightening your awareness of sounds in your environment.

SOUND-ALIKES *(Listening)*

ANSWERS ON PAGE 150

▶ *This puzzle exercises your ability to imagine and identify the sources of sounds.*

In the 1950s a film editor at Universal Pictures named Jack Foley came up with the idea of assembling a studio where he could create live sound effects for movies using simple and readily available sources. If the script called for sliding a block of ice

A

▶ ▶ ▶

across a floor, Foley created that sound by rolling a bowling ball instead.

Foley's genius at creating special sound effects inspired others to expand on his work. Foley artists—also called Foley walkers, because walking is one of the most common sound effects—are often trained dancers who are attracted to the field because of their well-developed sense of timing. They also must possess a sharp ear for sounds involving everyday objects as well as have the ability to employ those sounds to create special effects.

For instance, the sound of laser blasts in *Star Wars* was created by hitting with a hammer a tension wire supporting an antenna, then processing the sound in the studio. A sound produced by crunching newspaper is often used to produce the effect of . . . Try it yourself. While listening carefully, quickly and loudly roll a newspaper into a ball. Sounds just like . . . what? (The answer is in the puzzle on the next page.)

Sometimes substitute sounds are even more convincing than the original: the real sound of a fist striking a face isn't actually all that dramatic, but the Foley sound substitute really makes you squirm. And while some sound effects are straightforward—a door sounds like a door—others are less obvious. For instance, imagine that the script calls for you to step into a boat. Without bringing a boat into the studio, how would you create that sound? How would you create the sound of a horse's hooves in the absence of a horse?

On the next page is a puzzle in which you will match the sound effect you're trying to produce with the source that will help you create that sound. In some cases the source material is easily available and you can listen to the sound made by it (crunched newspapers). In other cases you'll have to imagine

▶ ▶ ▶

▶ ▶ ▶

what the source sounds like (e.g., walking on cornstarch). Thus, the puzzle tests your ability to really listen to the sound made by everyday objects while, at the same time, imaginatively using those sounds to create special effects.

Thanks to veteran Foley artist Allison Moore, who helped me assemble the lists of sound effects and their sources.

SOUND EFFECT	SOUND SOURCE
____ 1. Block of ice sliding on floor	a. Balloon
____ 2. Bone crushing	b. Pen caps floating in glass
____ 3. Body stabbing	c. Luggage cart
____ 4. Horse's hooves	d. Creaky floor
____ 5. Brain surgery	e. Knife in watermelon
____ 6. Dog collar	f. Bowling ball on floor
____ 7. Clothes rubbing against oneself	g. Newspaper being crunched up
____ 8. Bicycle	h. Squeaky chair
____ 9. Rubber gloves	i. Wet chamois cloth
____ 10. Car suspension	j. Celery
____ 11. Boat	k. Set of keys
____ 12. Fire	l. Leather purse
____ 13. Ice cubes	m. Pillowcase
____ 14. Car seat	n. Coconut shells
____ 15. Walking on leaves	o. Walking on grass mat
____ 16. Walking on snow	p. Walking on cornstarch
____ 17. Walking on grass	q. Walking on quarter-inch recording tape

Y

ANSWERS TO SOUND-ALIKES

1. Block of ice sliding on floor	f. Bowling ball on floor
2. Bone crushing	j. Celery
3. Body stabbing	e. Knife in watermelon
4. Horse's hooves	n. Coconut shells
5. Brain surgery	i. Wet chamois cloth
6. Dog collar	k. Set of keys
7. Clothes rubbing against oneself	m. Pillowcase
8. Bicycle	c. Luggage cart
9. Rubber gloves	a. Balloon
10. Car suspension	h. Squeaky chair
11. Boat	d. Creaky floor
12. Fire	g. Newspaper being crunched up
13. Ice cubes	b. Pen caps floating in glass
14. Car seat	l. Leather purse
15. Walking on leaves	q. Walking on quarter-inch recording tape
16. Walking on snow	p. Walking on cornstarch
17. Walking on grass	o. Walking on grass mat

REFLECTION. Which sounds were easy to identify? Which were hard? Watch a television program and notice the sound effects. Which sounds were artificially enhanced?

Here is one further sound awareness game to play when you watch television. Watch a drama program or movie. Notice those occasions when you hear something before you see it. For instance, you might hear a telephone ring off screen or hear the voice of a character before he enters the scene. Which television programs use this device the most? Which do not?

In order to correctly solve the above puzzle, you had to remember the sound effect that you were trying to match to a sound source. You did this by retaining the sound effect in your working memory while you simultaneously imagined each of the possible sound sources. In many cases this was difficult, because, for example, you probably have never paid the sources of these sounds (the crunch of celery) much conscious attention or attempted them yourself (walking on quarter-inch recording tape).

Cocktail-Party Effect

Listening is not a passive process. What we hear depends very much on what we intend to hear. While engaged in a conversation in a noisy, crowded room, we have no difficulty focusing our auditory attention on what our companion is saying while ignoring other conversations within earshot. Psychologists consider this "cocktail-party effect" as an auditory version of the figure-ground phenomenon. The sound we're paying attention to is the figure, while the ground consists of any other surrounding sound. And just as with the visual figure-ground situations (ambiguous figures such as those in chapter 13, "Illusions"), only a limited number of background elements can be processed simultaneously. Two additional conversations seems to be the maximum that most people can attend to. If we try to "listen in" on more conversational exchanges, we find ourselves unable keep up our part of the conversation we're engaged in. Typically, we have to fumble for a response because we missed what our companion has just said to us. But however many conversations we're trying to monitor, our auditory attention is automatically captured if something that is said nearby engages our attention. Our name is the most powerful attention grabber.

In order to test for selective auditory attention, scientists use what's called the dichotic listening technique. Volunteer subjects sit in a soundproof studio and listen with headphones to different words delivered to each ear. The subjects are instructed to listen to the words from one headphone (the attended message) and ignore those from the other headphone (the unattended message). After a few trials the instructions are reversed and attention is paid to sounds delivered to the opposite ear.

While the listener can usually repeat out loud the word transmitted over the channel she is attending to, little or nothing can be repeated from the unattended source. Even something as drastic as a change in the language spoken over the unattended channel will usually go unnoticed. But there is one notable exception: if the subject's name or some other emotionally significant word is transmitted over the unattended channel, the focus of attention may shift to the formerly unattended channel—a laboratory version of the cocktail-party effect. In addition to the shift in attention, slight changes in heart rate or other autonomic system responses may be triggered. This effect is especially likely if the emotional word is played through the left ear channel, which feeds principally to the right hemisphere where the emotional valence of words is processed.

In general, certain sounds are heard best when delivered to one specific ear. That's because each ear delivers its acoustic message primarily to the opposite cerebral hemisphere, with each hemisphere specialized to process different sounds. For example, most people hear words better through the right ear rather than the left (assuming they are not hard of hearing but instead have normal hearing in each ear). Pitch recognition or discrimination, in contrast, is best processed when heard by the left ear, which transmits primarily to the right hemisphere. Such findings are consistent with what we know about the two hemispheres.

The left hemisphere is specialized for speech and language. The right hemisphere can process language, too, but it isn't as good at it as the left hemisphere. Instead, the right hemisphere is specialized for all sounds other than speech or language, with one notable exception. Prosody, those everyday variations and nuances in inflection, rhythm, and stress that impart emotional meaning to words, is processed by the right hemisphere. This has important implications.

When we listen to someone talk, we listen not only to the words that are spoken but also to the speaker's prosody. And sometimes an aspect of prosody—a speaker's tone of voice, for instance—communicates more than the words themselves.

Neuroscientists learned about the emotional processing advantage for words of the right hemisphere from the study of brain-injured patients. Damage to the right hemisphere leads to difficulties in perceiving or expressing one or more of the elements of prosody. This can result in a conflict between the literal and the intended meaning of a sentence: "He is a real genius" can express admiration (a genuine genius) or ironic contempt (a nitwit), depending on the inflection placed on the word *genius*. A person with damage in the right hemisphere—in an area roughly corresponding to the speech areas in the left hemisphere—has great difficulty either speaking or understanding such an emotionally nuanced sentence. Interestingly, such people experience normal human emotions, although they have difficulty expressing these emotions or appreciating them in others.

Nor is this right-hemisphere dominance for the emotional aspect of speech restricted to people with brain damage. If you listen through earphones to someone speaking, you'll detect emotional nuances in his voice far better through the left earphone, which conducts the speech sounds to the right hemisphere. As a practical application of this experimental finding, if you want to detect emotional responses when you're talking to somebody on the telephone, hold the phone to your left ear rather than the right. In other conversational situations, make an audio recording and listen to it later through an earphone placed in your left ear. Try to ignore the content of the conversation and listen for changes in prosody.

On occasion, hearing and vision can come into conflict. What we see exerts a powerful influence on what we hear. An excellent demonstration of this is the so-called McGurk effect, first described more than thirty years ago by psychologists Harry McGurk and John MacDonald. In this effect, if we watch a video of someone's lips, teeth, and tongue forming one sound, /ga/, while listening on an audio channel to the sound /ba/, most of us will report hearing /da/, which is a fusion of /ba/ and /ga/. This fused response disappears if we look away or simply close our eyes and just listen. Then we hear the correct response, /ga/. You can experience this visual-auditory illusion for yourself on YouTube: "Hear with Your Eyes: The McGurk Effect," created by Hackszine.

Another example of sound-vision conflict can be found at www .cns.atr.jp/~kmtn/soundinducedillusoryFlash2/index.html, where you will perceive a single flash of light on a computer screen as two flashes if that single flash occurs simultaneously with two very short beeps.

Here is a puzzle that focuses you on the sounds of spoken language:

BACK-WORDS (*Listening*)

ANSWERS ON PAGE 155

▶ *This puzzle exercises your ability to notice and imagine the sounds in speech.*

The Speech Dissector exhibit at the Exploratorium science museum in San Francisco lets visitors experiment with recordings of their voices. A particularly interesting game is to record a word and play the recording backward. The results are often surprising. The word *we* (oo-ee), for example, backward becomes *you* (ee-oo). If you play each of the words below backward, what other word does it become? Can you find other backward word pairs?

A. no
B. yaw
C. yes
D. lane
E. view
F. spin
G. weak
H. shall
I. funny
J. ominous

ANSWERS TO BACK-WORDS

A. no—one   F. spin—nips
B. yaw—eye   G. weak—queue
C. yes—say   H. shall—lash
D. lane—nail   I. funny—enough
E. view—weave   J. ominous—cinema

REFLECTION. Which words were easy to reverse? Which were hard? How did you figure them out? Try reversing the sound of your first name or the name of a friend.

The Sounds of Music

Music is an important part of most people's lives. You don't have to be a musician to love music.

Oliver Sacks, in his book *Musicophilia: Tales of Music and the Brain,* writes about patients with Parkinson's disease who, over long spells, were unable to move or to talk. But when music was played for them, they were magically released from their bondage. They couldn't walk but they could dance. They couldn't speak but they could sing. A similar distinction is found among people who stutter. The country-western singer Mel Tillis sings without any hint of a stutter, but he rarely gives interviews, because those same words that he can sing flawlessly are stuttered when spoken in conversation.

But not all of us are musicians. Here is a musical exercise that you don't have to be a musician to enjoy. It focuses on the most cognitively important aspects of music: rhythm, pattern, and the interactive patterning that occurs whenever two or more musicians play together.

F

CLAP YOUR NAME *(Listening)*

▶ *This game strengthens your ability to play and listen to rhythms.*

(Adapted from Math Dance with Dr. Schaffer and Mr. Stern *by Schaffer, Stern, and Kim)*

Sound out the letters of your first name by clapping your hands for a vowel and slapping your thighs for a consonant. For example, MARIA would be played

Slap-clap-slap-clap-clap

Learn to play your name pattern without pauses so every clap and slap takes the same amount of time.

Now play your name three times in a row, without pausing at the end of your name. Make sure every clap or slap lasts the same amount of time. Many people naturally want to pause at the end of their name:

M-A-R-I-A——M-A-R-I-A——M-A-R-I-A

For this game we want you to go right back to the beginning:

M-A-R-I-A-M-A-R-I-A-M-A-R-I-A

Also, accent the first letter of your name by playing it a bit louder. This makes it easier for you to hear the beginning of the pattern. For example, accenting the M in MARIA creates a five-beat pattern:

M-A-R-I-A-**M**-A-R-I-A-**M**-A-R-I-A

Again, play your name so that every beat takes the same amount of time. Practice until the pattern becomes easy for you and you can play it softly with relaxed hands. If you prefer, you can sub-

▶ ▶ ▶

▶ ▶ ▶

stitute other sounds for the clap and the slap. For instance, you could tap the table and stamp your foot.

Pair up with someone whose name is not the same number of letters as yours. Sit facing each other and, on the count of three, play your names at the same time, and at the same speed. Keep repeating your names at least five times. Try not to get confused by listening too closely (at first!) to what the other person is doing. Do not speed up or slow down: a consistent beat allows each of you to continue playing without getting confused. For instance, playing MARK and ZOE together looks like this:

M	A	R	K	M	A	R	K	M	A	R	K	M
Z	O	E	Z	O	E	Z	O	E	Z	O	E	Z

Practice until both of you are good at clapping your names and can repeatedly clap them without distracting each other. As you get better, your hands will play the pattern automatically, and you can enjoy listening to the interplay of rhythms that you are creating together. Skilled musicians in groups learn to listen carefully to other instruments while they are playing their own instruments.

This game involves several beneficial brain effects. First, the clapping and slapping pattern has to be learned so that it can be performed seamlessly and without error. Depending on the number of letters in your name, this will require varying periods of time to establish a flawless rhythm. After the pattern has been established in the circuitry of the motor neurons and you've mastered the solitary pursuit of clapping your name while alone, begin clapping your name in the company of others while they simultaneously clap their names. At first you'll find this distracting and you'll lose your rhythm. Return to solitary clapping once again and, after additional practice, return to clapping with one, then two, and then as many people as you can get to attempt this challenging exercise in concentration and motor sequencing. After you've become proficient in group name clapping, ask one of the other participants to clap out his middle name. See how many names you can come up with that correspond to the clapping pattern. Speak the names and learn if you've included the person's middle name on your list.

Clapping your name also relates to the subject of our next chapter: the learning of motor skills.

7

MOTOR SKILL LEARNING:
OF MENTAL MAPS AND PICKPOCKETS

MOTOR SKILL LEARNING involves the activation of several brain areas: the posterior parietal cortex (which appreciates, for instance, the physical characteristics of a puzzle), the supplementary motor area (which plans the physical responses to the puzzle's challenge), the cingulate cortex (which provides insightful problem solutions), and the cerebellum (which programs all of the steps that will be involved). Since motor skill learning involves actually doing something (usually with one's hands, as in the Clap Your Name exercise described in the last chapter), it's a great way to enhance eye-brain-hand coordination. This is especially important because the maintenance of skilled finger movement is associated with prolonged longevity.

To increase motor skill learning, try games and exercises like Jenga, pickup sticks, and model building: each is great for developing small, fine-muscle coordination. Exercises involving the body's larger muscles include anything that challenges your balance and coordination: dancing, tai chi, yoga, and the following two-handed body coordination game.

TRIANGLE VS. SQUARE *(Motor Skills)*

▶ *This activity strengthens physical coordination.*

Here is a challenging coordination game that requires your two hands to do two different things.

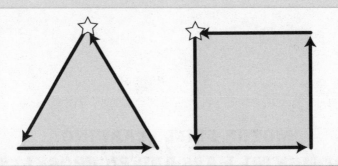

Place the tips of your index fingers on the stars above, with your left index finger on the left star and your right index finger on the right star.

Move both hands simultaneously counterclockwise around the edges of the shapes till your index fingers reach the next corners. Keep going, moving both hands counterclockwise around the shapes, one corner at a time. Since the triangle has three sides and the square four sides, the two hands will not return to the beginning at the same time until both have gone around several times. With practice, this skill becomes easier.

I find that I can do this best if I imagine the muscular sensation of how each hand moves to the next corner, mentally rehearsing the coordinated motion before I actually do it. Instead of performing two unrelated motions, I have the sensation of performing one coordinated motion, a mental technique similar to the chunking technique mentioned in chapter 2.

By the way, I have always enjoyed coordination challenges like juggling. I spent many years learning to play the piano,

> ▶ ▶ ▶
>
> which taught me how to make my two hands do two separate but related things simultaneously. I particularly enjoy playing Chopin études with multiple rhythms, and Bach fugues with multiple melodic lines.

The brain areas used in skill learning also differ from those areas used for so-called episodic memory, i.e., memory for personally experienced events such as last weekend's bicycle trip or office picnic. While episodic memory depends on the temporal lobes (especially the medial parts), skill learning (riding your bicycle or preparing the dish you brought to the picnic) depends on the cooperative action of the cerebellum (the structure at the very back of the brain), the cerebral cortex, and the basal ganglia.

Despite its off-putting name, the basal ganglia are simple to conceptualize: a series of *ganglia* (clusters of neurons) located at the *base* of the cerebral hemispheres. The basal ganglia are important for controlling the speed, direction, and amplitude of all of our movements.

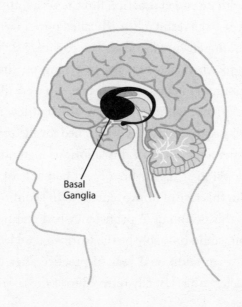

Basal
Ganglia

Here's a short summary of the actions of the basal ganglia, based on many experiments.

Whenever we learn a new movement sequence (e.g., coordinating the clutch and gearshift of a manual transmission), one or more components of the basal ganglia are activated. This activation links our perceptual-sensory processes (correctly appraising the location of the gearshift and the clutch) with our motor responses (efficiently shifting forward in sequence starting from first gear). An important part of this involves suppressing undesirable movements and carrying out only those movements needed for efficient gear shifting. Parkinson's patients and other people with diseases of the basal ganglia have great difficulty doing this because they cannot make the transition from correctly *describing* how to operate a manual transmission to actually doing it.

In contrast to motor skills like learning the mechanics of driving, cognitive skills involve *mental* activities aimed at solving problems. Preparing tax returns, keeping to a budget, taking a standardized test, and successfully solving a puzzle are examples of cognitive skills. As a rule, cognitive skills, like motor skills, improve in direct proportion to the time devoted to them. So the amount of time needed to perform a skill decreases with extended practice. But there's a limit to the pace of this improvement—the result of the so-called power law of learning— that applies to both perceptual-motor and cognitive skills.

Think back to when you first began typing on the keyboard of your PC. (I'm assuming in this example that you were like me and had no previous experience with typing.) For the first few months it was strictly hunt-and-peck as you visually searched for the correct letters on the keyboard. After a year or so of this laborious method, you were less dependent on looking at the keyboard and had at least doubled your typing speed. But this rate of improvement didn't continue at the same pace. Indeed, if this doubling in proficiency had continued year after year, by now you would probably be the world-record holder for speed typing. Instead, each additional year of practice after the first led to smaller and smaller gains. Usually referred to as the law of diminishing returns, this effect is more formally referred to by psychologists as the power law of learning.

One way of overcoming the power law is to take advantage of feedback about your performance. But what kind of feedback and how often should it be provided? The secret to improvement in skilled performance is to correlate a particular skill with the kind of feedback that leads to maximal improvement in that skill. For example, watch videos of yourself in action. Athletes do this all the time. But there's a catch here: in the long run, too-frequent feedback degrades rather than improves skill learning. If this seems counterintuitive, imagine the effect on your golf game if your golf instructor accompanied you and commented on your every move each time you went out on the course. Soon you would be so overloaded and stressed that your performance would deteriorate rather than improve. Like medicine, feedback must be taken at the correct dose and frequency. Infrequent feedback, in contrast, isn't as helpful in the early stages of learning but, if continued, leads to improved performance in the long term. In this golf example, you would take lessons on an as-needed basis, with each lesson followed by deliberate solitary practice.

In support of these learning principles, my friend psychologist Mark Gluck reminded me of a classic experiment that involved postal workers trained to use a keyboard to control a letter-sorting machine. In the experiment, four groups of postal workers received instruction in the use of the keyboard. One group received instruction for one hour a day over a three-month period (spaced practice). The other three groups trained either two or four hours a day for a month (massed practice). Which of these groups do you think required the fewest number of practice hours to become skilled at using the keyboard? Those who trained for only one hour a day (spaced practice) required fewer hours of training than any other group in order to become proficient at using the keyboard. On the downside, the employees trained by spaced practice had to be trained over a longer period—three months instead of one month—in order to achieve their highest level of expertise. So if you want to learn a skill and retain your abilities, practice every day for short periods of time and continue practicing rather than practicing for many hours a day over a brief time span.

The effectiveness of practice also depends on whether the practice pattern remains the same from one practice session to another or whether it varies across practice sessions. Constant practice involves doing the same thing each time, while variable practice incorporates variations and fresh approaches. At high levels of skill learning, both types of practice are usually involved. Repeatedly shooting baskets from the foul line can help perfect your accuracy in putting the ball through the hoop, but if you want to be a world-class player, you have to practice from just about every position on the court.

In summary, think of learning in the brain as relying on two independent circuits. One circuit manages things, facts, specific memories, and experiences (episodic memory). The other circuitry manages processes and procedures—doing things. Failures in one circuit do not necessarily involve failures in the other. People who suffer from extreme memory failures for facts and personal experiences, for instance, can be taught to improve their performance on motor tasks. A classic example of this is the improvement with practice that can be brought about in mirror drawing: tracing figures or pathways seen in a mirror.

You can experience this improvement with practice in mirror drawing for yourself by printing out the path depicted in the figure on the opposite page and putting it on a desk in front of a small mirror. Place a pencil at Start and, by observing the movement of your hand in the mirror, move the pencil along the path to Finish. It's important that you don't have any opportunity to observe your hand's movement except in the mirror. The easiest way to avoid this is to hold your nonwriting hand over the drawing; that way you can't inadvertently look directly down at your hand making the drawing.

Unless you have a natural talent for mirror drawing (in which case you can probably mirror write as well), you'll find this exercise initially difficult. But with practice you'll improve. At no time, however, will you be able to improve anyone's performance at the task simply by describing to them how you learned to do it; in fact, you won't be able

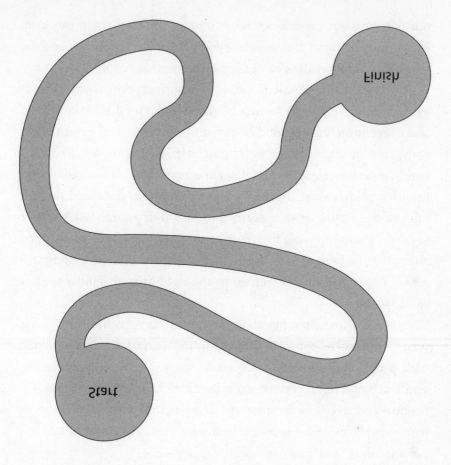

to describe the process to yourself, either, because skill learning isn't readily translatable into words. (That's why golf, tennis, and other sports are taught by lessons and not lectures.) Each person will have to practice it for himself until he learns to override the perceived position of his hand (proprioception) and concentrate on controlling the movement of the hand strictly by observing its motion in the mirror.

Mirror drawing is difficult because, at least initially, we're unable to reconcile the mirror-reversed pattern with the novel hand actions required to duplicate that pattern. Ordinarily we look at the looping pattern on the paper and simply draw what we *see*. Our eyes and the drawing hand effortlessly work together. But when we can't look directly at the patterns, as in the mirror-drawing exercise, we experience a conflict between what our eyes and hand are "telling" us. Ordinarily,

A

our eyes and our proprioceptors (the sensory receptors in our skin, joints, and ligaments that sense the forces acting on our body and the movements and position of our limbs in space) are coordinated.

To get an intuitive understanding of proprioception, try this simple exercise: Raise your right hand to about shoulder level, close your eyes, and concentrate for a second or two on your sense of the position of your hand in space. Ignore everything else. That inner sense of your hand's position in space is about as close as you can come to conscious proprioception (which, by the way, is an oxymoron, since by definition our proprioceptive sense operates outside of our awareness). In order to eclipse that proprioceptive sense of your body in space, simply open your eyes and look at your hand. At that moment, the proprioceptive sense of your hand is overridden by the sight of your hand—or so it would seem.

When you first attempt a mirror drawing, seeing comes into conflict with the touch receptors embedded in the skin of your hand, along with proprioceptors within the muscles, tendons, and joints in the arm and hand. But if you persevere in mirror drawing, your brain will readjust and reconcile the mirrored image with the sensations coming from your arm and hand. After enough practice, you can become accomplished at mirror drawing.

MIRROR WRITING *(Motor Skills)*

▶ *This game strengthens physical coordination.*

Here is a game that is closely related to mirror drawing. For this game you do not need a mirror, just a big sheet of paper and two pens, one in each hand. For this to work, the paper should not move as you draw. A large drawing pad works best, but you can also use a writing pad or a piece of paper taped to the table.

▶ ▶ ▶

If you have a whiteboard, you can also use that. Stand facing the board with a whiteboard marker in each hand.

Warm up by pretending to conduct an orchestra by waving both your hands. Don't write yet. Move your hands in opposite directions, making all different sorts of patterns. You can move your hands in arcs, circles, zigzags—anything, as long as your hands move in opposite directions.

Now let your pens drag along the paper as you continue to move your hands in opposite directions. Don't be concerned 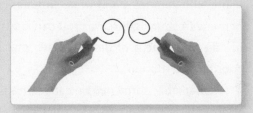 about what you are drawing. Instead, focus on the muscular sensation of moving your hands.

Now, if you are right-handed, put both hands in the center of the page. If you are left-handed, put both hands at the far edges of the page. Then, with both hands simultaneously moving in opposite directions, write your first name. Your dominant hand will write forward and your other hand will write backward. Some people find cursive easier; others prefer printing.

This may seem impossible, but don't worry, it's easier than it sounds, because moving your arms in opposite (mirror-image) directions is something your brain already knows how to do. In fact, in order to make it even easier, do not look at your hands. Just feel the movement—what neuroscientists call proprioception—and let your hands move naturally.

When you're done, pick up the paper and look through the back to check your work. If you're drawing on a whiteboard, look at the board in a mirror to check your work. Which letters were

▶ ▶ ▶

▶ ▶ ▶

easy to draw? Which were hard? Try it again and see if you can do better.

With practice, you'll be able to do this quickly and with little conscious effort. Because it seems impossible when they first hear it described, people are surprised and delighted when they are able to do it, which is typically very quickly.

A much harder version of this game is to write your first name with your left hand while simultaneously writing your last name with your right hand. You can learn to do this by rehearsing the movement of drawing each pair of strokes as a single coordinated motion.

Remarkably, some people can learn to write completely independently with both hands. One performer listed in *Ripley's Believe It or Not!* learned to write four different words simultaneously with her two hands and her two feet.

Mirror drawing and the Displaced Hand game described on page 174 illustrate an important point about our sensations. While vision is our dominant sense (more neurons are devoted to seeing than to any of the other senses), we're more reliant on proprioception than we usually realize. Blindfold yourself for fifteen minutes and try doing some routine (and safe) activities like making your bed, dressing yourself (put aside the clothes prior to putting on the blindfold), taking a shower, and brushing your teeth. You'll be surprised how well touch and kinesthesia can substitute for vision when you do these things. In fact, vision doesn't add anything at all. To confirm this for yourself, remove the blindfold, undress, and then dress yourself again, only this time deliberately focus your vision on the act of buttoning buttons, zipping zippers, and tying shoelaces. You'll find that you don't accomplish these activities any better by looking at your hands while you do them. In some cases, such as tying shoelaces, too much visual focus actually interferes with getting the job done. And there's good reason for this;

touch and proprioception are specialized to provide direct physical contact, while sight and hearing provide indirect perceptions of distant events and objects. Since we're in direct physical contact with our clothes, we don't have to look at our buttons or zippers except when something goes wrong (a button snaps or a zipper gets stuck).

A dramatic example of the perils that can ensue with the loss of proprioception is vividly described in neurologist Jonathan Cole's book *Pride and a Daily Marathon*. At age nineteen, Cole's patient, Ian, contracted a viral infection that destroyed the connection between the brain and the nerves controlling proprioception. Although Ian is not paralyzed—the motor nerves were spared—he is still unable to walk, because in the absence of input from the kinesthetic receptors in his limbs, he can't determine the position of his arms and legs in space without looking at them. Over several years Ian learned to move about again by substituting vision for his proprioceptive sense. By looking at his hands and feet, he can move in a seemingly normal manner. But if he's deprived of light, he can't move or even remain standing because of his lack of an internal sense of the position of his arms and legs. On one occasion Ian collapsed to the floor when the lights in the elevator he was riding in suddenly went out.

Ian's experience suggests that we underestimate the importance of proprioception as a means of helping us localize ourselves in space. But proprioception is also crucial in regard to our spatial sense. For instance, imagine yourself driving from home to office or some other familiar destination. Since you've made the trip many times you're not aware of the many occasions during your drive when you turn the wheel in a direction *opposite* to your final destination (driving up a curving ramp to get on the freeway, for example). If the highway curves are gentle enough, you may not even be aware of many mild angular departures from the direction of your goal. Even detours and traffic-related deviations from your usual route will have little effect on your getting to your destination. That's because, after having driven this route many times before, you have unconsciously incorporated into your brain a spatial map of your trip. And that map is based not just on the landmarks you've seen during previous trips but also on the stimulation of your proprioceptors each time you turn the wheel, accelerate, and apply your brakes. These spatial skills play a large role in the construction of the spatial map—which, incidentally, doesn't necessarily translate into facility in conveying that spatial knowledge to other people.

For instance, you may not be able to draw a route that you take on a regular basis or give directions to other people so that they can take the same route. Furthermore, it may take a long time to establish this spatial map and require your active participation: when you first learned the route you never felt completely confident until you drove it yourself. Sitting in the passenger seat while somebody else did the driving just wasn't good enough. At some point you had to get behind the wheel and use the feedback from your own proprioceptors to form your internal map.

Spatial ability, whether acquired by driving a car or walking a familiar route, involves the incorporation of proprioceptive movements into the formation of a map within the brain. This mental map doesn't have to be precise since, as geographer Yi-Fu Tuan pointed out more than thirty years ago in his book *Space and Place: The Perspective of Experience*: "Precision is not required in the practical business of moving about. A person needs only to have a general sense of direction to the goal, and to know what to do next on each segment of the journey." The driving-to-work example mentioned above illustrates Tuan's point well.

Experiments in maze walking provide additional evidence. Learning to wend one's way while blindfolded through a maze demands the integration of proprioceptive patterns. But even after these patterns have been integrated into the brain, they can't usually be recalled. After successfully learning to navigate the maze, most people can't walk the same pattern on an open floor. Nor can they provide a reasonably accurate drawing of the maze as illustrated in the figures on page 171. In fact, few people who have successfully learned to navigate a maze while blindfolded can even provide a verbal account of the sequence of their turns as either "right" or "left." If they attempt to do so, they usually give up in the midst of their recitation and say something like "I don't know what comes next. I have to be there before I can tell you."

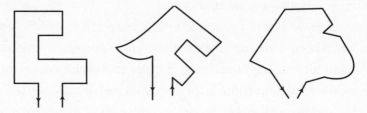

The Skill of the Pickpocket

Touch exerts a particularly powerful effect on our attention. If you are told to close your eyes and expect to feel a light touch applied at a particular location on your body (usually the face or the hand), you'll be able to focus your attention well enough to precisely identify the spot where you've been touched. Interestingly, if you are touched in two places (both the face and the hand), your brain tends tend ignore the touch on the hand and report just the touch to the face. This so-called face-hand test is a standard neurological exercise for detecting subtle brain disorders. But even people with perfectly normal brains may "extinguish" the touch to the hand because they are so focused on the expectation of being touched at only one location. (The neurologist's command "Close your eyes and I will then lightly touch you. Please report where you feel me touching you" contributes to the mistaken notion that only one touch will be delivered.)

On occasion, the touch and proprioceptive sense can falsify the testimony of our eyes. In one classic demonstration of this dating from the 1980s, people holding a square looked at it through a distorting lens that made the square look like a rectangle. They immediately described the figure as a rectangle. In other experiments, the balance between proprioception and vision may shift according to circumstances. In one experiment, people looked at and felt different sandpaper surfaces that they had previously been told were the same. When they were asked about the proximity of the raised features on the sandpaper, they responded by looking; when asked about texture and roughness, touch became the dominant sense employed.

Whenever we switch from touch-proprioception to vision or vice versa, we leave ourselves open to misdirection, a form of deception that focuses our attention on one thing in order to distract our attention from another. Misdirection is best exemplified by the art of the magician and the pickpocket. Since I'm a member of the International Brotherhood of Magicians and, as a condition for membership, I agreed never to provide explanations for magic tricks to nonmagicians, let's concentrate on the art (and it *is* an art) of the pickpocket.

Before explaining how pickpockets practice their nefarious trade, I invite you to watch in action probably the most famous (infamous?) of contemporary pickpockets: Apollo Robbins. Link to the YouTube video "Apollo the Pickpocket" and watch at least a couple of times the six-second snippet of Apollo talking to a smiling couple (the third example in this 1 minute, 33 second video). In that brief interval, Apollo removes the man's watch without arousing any perception of the act on the part of the "mark." How did he do it? It's all there in plain view. Clue: Think about what I mentioned in the previous paragraph about switching from touch-proprioception to vision.

In the video, you see Apollo walking up to the couple and, with his left hand, touching the man's left shoulder. At the same time, his right hand is gently squeezing the man's left wrist (the one with the watch). He then misdirects the man's attention by pointing with his left arm downward as though looking at something. As the man follows Apollo's gaze his watch is pilfered.

Apollo's wizardry involves misdirecting the mark's attention from what's going on—the removal of the watch—which would ordinarily activate the touch and proprioceptive receptors of the man's wrist and arm. Apollo starts by moving in very close and slightly off to the side of his victim's direct vision. The friendly touch on the shoulder misdirects attention there while the other hand slightly squeezes the wrist on which the watch is still in place. This causes the touch receptors in the skin at that site to adapt, thus rendering them less sensitive to the subsequent light touches needed to open the buckle and slip the watch off. That light squeeze also leaves in its wake a sensory after-image, which creates the illusion that the watch is still on the wrist after its removal. Notice, too, that as Apollo snags the watch, he makes a sweeping movement with his left arm, followed by a straight motion directed downward toward something he appears to be looking at on the ground. The mark and his wife also look in the direction of Apollo's gaze—an additional misdirection, since the real action at this point involves touch and proprioception, not vision.

One more point: Apollo's initial sweeping movement, followed by the straight hand movement, activates two different visual control systems. The sweeping movement activates in the mark the eye's smooth pursuit system, which locks onto the curvilinear trajectory of Apollo's arm, thus diverting attention away from what's about to happen. The sudden shift from a curvilinear motion to a fast, downward-pointing motion then activates a second visual system that is involved in fast attentional shifts from one spatial location to another. During the duration of this rapid attentional shift, the mark's vision is temporarily suppressed as his eyes switch from smoothly following the curvilinear path of Apollo's arm movement to abruptly looking straight downward toward the ground, where Apollo appears to be staring at something.

Keep in mind that this is only a partial explanation of what's going on in this six-second gem illustrating the art of the pickpocket. You can learn more about some of the other elements, especially the importance of subtly entering the mark's personal space, from Apollo himself by linking to his explanation of the art of the pickpocket. You can find that by going to the sixteen-minute video "Apollo Robbins—Magic of Consciousness Symposium." This was part of Apollo's 2007 presentation at the Association for the Scientific Study of Consciousness held in (where else?) Las Vegas, Nevada.

When it comes to influencing our reactions, proprioception can on occasion trump vision, as you can see in another video on YouTube, "The

Rubber Hand Illusion," footage courtesy of the École Polytechnique Fédérale de Lausanne and narrated by Professor Olaf Blanke.

In the video, you will see a vivid demonstration and account of a young woman's sense of her hand being altered by something as simple as hiding it from sight under a table and stroking one of her fingers with a fine brush. At the same time, she watches an identical pattern of stroking applied to a rubber hand sitting on the table before her. After a few minutes of observing this, the touch sensation from the brush is displaced from her real hand under the table to the rubber hand sitting before her on the table. Even though she knows the rubber hand isn't real, she nonetheless experiences the stroking as if it emanates from there.

DISPLACED HAND (*Motor Skills, Visual Thinking*)

▶ *This activity challenges coordination between your proprioceptive and visual senses.*

Here's a version of the proprioceptive/vision illusion you can do at home. You'll need a door that has a large mirror on it, large enough that it comes close to the edge of the door at the height of your head.

Stand facing the edge of the door with your hands holding the doorknobs (if there are no knobs, just touch the opposite sides of the door). Without moving your feet, tilt your head a little so your head is com-

pletely on the mirrored side of the door. For the sake of discussion, let's say you tilted your head to the right.

Extend your arms out to opposite sides of the door and move your hands in opposite directions. Try tapping the door

▶ ▶ ▶

▶ ▶ ▶

with both hands, moving your hands in circles in opposite directions and twisting your wrists so both hands are palms up, then palms down. Moving your hands in opposite directions is a very natural motion, so this will feel easy and comfortable.

Since you are looking into a mirror and your hands are moving in mirror symmetry, you will have the odd sensation that you are looking at your left hand in the mirror, even though what you see is a reflection of your right hand. Keep moving your hands in opposite directions until the illusion of seeing your hand seems completely real.

Now suddenly, without thinking about it, move one hand up and the other down. If you can catch yourself off guard, so to speak, you will have the very weird experience of having your visual image of your left hand not match your proprioceptive image of your left hand.

If the powerful influence of the proprioceptive sense is reduced, the normal integration between visual and proprioceptive maps in the brain disappears and vision becomes the dominant influence. Mirror drawing, for example, improves when a harmless pulse of magnetic stimulation is applied to the part of the brain responsible for proprioception from the hand. Ordinarily, as previously mentioned, mirror drawing is difficult because visual feedback from the mirror conflicts with the proprioceptive sense of the hand's position. The brief magnetic pulse temporarily reduces the excitability of those cortical areas responsible for the proprioceptive sense of the hand's position. Thus, freed by the magnetic pulse from the interfering effect of the proprioceptive sense, visual feedback from the mirror then becomes sufficient for the person to mirror draw. The same thing happens in people like Ian (described earlier) who lack their proprioceptive sense. They learn to mirror draw four times as fast as people with normal proprioception. In the absence of proprioception, the mirror drawing is a simple tracking task.

SCOTT That's exactly what I was referring to when I mentioned above that you can improve your performance in the two-handed simultaneous-writing-of-your-name game by not looking at your hands. That way, the proprioceptive sense isn't interfered with by what you're seeing, just as you get better at mirror drawing by concealing your hands.

RICHARD On occasion, though, I find that I have to look at my hands in order for me to get the hang of how to do one of your puzzles or exercises. The plate exercise below is a good example. Please give a description of it and each reader can judge for herself how she went about doing it.

OPPOSITES CIRCLES *(Motor Skills)*

▶ *This activity exercises physical coordination.*

Pretend you are holding a large plate between your hands, so that you are looking at the plate edge-on. Remove your hands and let the imaginary plate stay suspended in the air. Then place both forefingers at the top of the rim of the plate. With both hands simultaneously, trace around the edge of the plate in opposite directions, in a continuous circular motion. When you get to the bottom of the plate, your fingers will touch and pass each other. Keep going. Your fingers will keep passing each other at the top and at the bottom of the plate. This motion is quite confusing at first but becomes easier with practice.

Try this exercise with your eyes closed. Can you make your fingers touch as they pass each other at the top and bottom of the circle? Then try it with your eyes open and looking at your hands. Does looking at your hands make this exercise easier or harder?

▶ ▶ ▶

> ▶ ▶ ▶
>
> To make the challenge harder, move your hands at different speeds—for instance, one hand makes one revolution in the time the other hand makes two—or make one circle smaller than the other.

RICHARD At first I found I couldn't do this exercise because both of my hands, when reaching the bottom of the imaginary plate, tended to go off on separate tangents instead of continuing on their clockwise and counterclockwise circular paths. Then I got a real plate, held it so that was I was looking at it edge-on, and then simply rested the plate flat on the table and repetitively performed the circular motions with my hands. Now I can do the mental exercise quite easily.

Learning Navigational Skills

The study of London taxi drivers mentioned in the Introduction was one of a series of experiments identifying the hippocampus as important in navigation. Rats provided the initial insight with the discovery of place cells within their brains, which became sequentially active as the animals moved from one location to another. Change the location, even slightly, and different cells in the rat brain begin to fire. A similar process occurs in our own brains. As we drive from home to work, a series of cells within our hippocampus turn on as we approach a major intersection or pass a familiar store. Encountering each of these landmarks is associated with the firing of different hippocampal cells. This process breaks down if we become lost. At such times our place cells no longer fire because we don't recognize where we are.

As we get older, this loss of navigational skills can occur even in situations where we recognize our current location but aren't entirely certain how to navigate from there to our destination. According to

Scott Moffat of Wayne State University, who has studied navigation in older people as they play a virtual environment navigation game, the older person recognizes all of the relevant landmarks but has difficulty associating these landmarks with the correct direction. One of the reasons for this failure, according to Moffat, is the older brain's difficulty in filtering or disregarding irrelevant information. Incidentally, Moffat's studies, along with those of other researchers on spatial navigation, involved normally aging people with brains that were working just fine for their age.

Since navigational skills decrease with age, it's helpful to bolster them with specific exercises and puzzles. A simple and readily available way of doing this is to compete against your GPS. Although overdependence on a GPS can lead to a deterioration of one's sense of spatial orientation, it can also enhance that sense when properly used. For instance, mentally envision every relevant navigational detail of your daily commute from your home to the office. Now write them out or dictate them into a voice recorder. Later, when driving, play the recording and see if you've missed any of the key details. Or the next time you make the trip, use your GPS as a check on your internal navigation map. As you approach a key navigational landmark, name it before the GPS does.

As you'll discover, recognition of key navigational sites isn't necessarily linked with an ability to name those sites. You won't be able to name many of the streets that you can see in your "mind's eye," because you never needed to process that information: you simply knew where to turn and did so when you reached that street. Often this disparity between intact navigational skills and naming failures becomes obvious only if we're asked to give someone directions. This is one of the reasons why some people when asked for directions resort to creating a diagram: they "know" the correct directions but can't name the relevant navigational landmarks. Others who are weak on navigational skills work from a linguistic rather than a spatial map. But if they lose the detailed verbal directions they're working from, they haven't a clue how to proceed. Totally lost, they have to pull over and ask for directions (best given by someone with a similar preference for language rather than spatial maps).

Try this exercise whenever you use your GPS: Study a map of where you're going prior to starting out and then pit your hippocampal place cells against the GPS.

Practice can enhance our navigational skills. It can also delay the age-associated fall-off in navigational skills noted by Professor Scott Moffat. Here is a practical exercise that you can do at home:

NAVIGATE YOUR HOUSE *(Motor Skills)*

▶ *This puzzle exercises your ability to imagine paths you travel in space.*

Choose two places in your house that are as far apart as possible. Without moving, close your eyes and imagine walking from one place to the other. Notice your path. Then write down instructions for how to walk the path.

Now walk the path holding the instructions. How accurate were your instructions? What sorts of mistakes did you make? Revise your instructions so they are more accurate.

For this next part you'll need the help of a friend. Take him to the starting place blindfolded or with his eyes closed. Read the instructions to get your friend to the other place.

Was he able to get to the destination? What sorts of mistakes did you make in your written instructions? Revise your instructions so they are more accurate. What sorts of instructions were particularly helpful? Did your instructions include easily identified landmarks? Sounds? Textures to be felt by the feet? Warnings about dangers to avoid?

R

8

TIME:
CLOCK TIME VS. BRAIN TIME

AT FIRST GLANCE, our appreciation of time seems dependent on words. We speak of time in terms of seconds, hours, decades, and centuries. But this reliance on spoken and written symbols isn't always true. Think back to the last time you judged whether it was safe to change lanes in rapidly moving traffic, or the occasion when you ducked instantaneously in order to avoid being hit by a baseball rapidly headed your way while you were in the bleachers at a ball game. Neither of these time estimations had anything to do with language. Rather, your brain automatically sized up these situations and fired the impulses needed for instantaneous reaction. And sometimes this internal sense of what psychologist's term *interval timing* is wrong: perhaps your lane change was too tardy and, were it not for the other driver braking hard, you would have been in an accident; or you didn't move fast enough to avoid the ball but, luckily, another other fan caught it milliseconds before it hit you. In instances like this, real time and subjective time can be out of synch—proof that one is not the exact correlate of the other.

A similar time discrepancy between clock time and subjective time can be demonstrated in the laboratory. If I ask you to press a button and hold it without counting for exactly 5 seconds, your response may vary by as much as 20 percent from one testing to another. If I ask you to hold the button for 50 seconds you will, on average, do so within a range of 40 of 60 seconds—again a variation of 20 percent. Our subjective sense of the passage of time also varies according to circumstances. Drugs can skew our sense of objective time by altering the balance of neurotransmitters (especially dopamine). Cocaine, caffeine, and nicotine increase brain dopamine and speed up our perception of time; sedatives like Valium and cannabis slow it down. But drugs aren't necessary to alter our sense of time. Depressed and bored people also complain that time seems endless.

Attention is the most important variable influencing our appreciation of the passage of time. The closer attention we pay to the passage of time, the slower it seems to go ("A watched pot never boils"). But there's a paradox here. While we may experience the passage of time as agonizingly slow when we're doing something like waiting for a pot to boil or sitting in a doctor's waiting room, we underestimate rather than overestimate the duration when we're later asked to estimate how much time actually passed.

In an experiment illustrating this, people watching an action movie experienced time passing faster than other people sitting in a waiting room. No surprise there. Yet when the two groups were asked to estimate how much time had actually passed in these two situations, the results were just the opposite: despite their subjective feeling that the time had passed quickly, the movie watchers later estimated the elapsed time at about 10 percent longer than the waiting room group.

The explanation for such paradoxical findings, according to John Wearden of Keele University in Staffordshire, who carried out the experiment, is that the two groups based their final time assessment on the amount of information their brains had processed. In the waiting room, not much was happening and time seemed to drag on. But looking back on the experience, the time span seemed shorter *because* not much had happened. For the viewers of the action movie, in contrast, time went quickly because a lot was happening. But on later recall, thanks to the sheer number of events that transpired in the movie, it seemed that more time had passed than actually had.

As a general rule, the number of things that happen during a specific interval, along with their complexity and emotional significance, tend to make the experience seem longer in retrospect than it actually was. If we're in an auto accident or get mugged, for instance, these events tend to be perceived as lasting much longer than they actually did. Although neuroscientists aren't entirely in agreement about the explanation for this time distortion effect, Peter U. Tse, a neuroscientist at Dartmouth College in Hanover, New Hampshire, and his associates provide, in the November 2005 issue of the *Journal of Neuroscience*, a most intuitively appealing explanation.

Tse proposes a simple "counter" model. Imagine that the brain estimates time in bits (units) based not on clock time but on its own rate of information processing. Thus, under most conditions, one bit of processed information corresponds to the passage of one unit of objective (clock) time. Now imagine the rate of processed information suddenly increasing to two or more bits per unit of clock time in an emergency situation such as suddenly focusing increased attention on the road in order to avoid a collision with another car. Now, because of the increased attention focused on this dangerous situation, the "counter" registers two bits rather than one per unit of objective time. "If the assessment of duration by the brain is the result of the output of such a counter, it would come to the wrong conclusion that more objective time had passed, creating the illusion that time and motion had slowed down," Tse commented during a mini-symposium, "Time and the Brain: How Subjective Time Relates to Neural Time."

Time estimates also vary as we get older. With less to do during retirement, more attention is likely to be paid to the passage of time, resulting in boredom and a sense of time "standing still." But later, when looking back on a series of such experiences, they are judged as happening more quickly because, in the absence of much taking place, the brain didn't process much information. As a result, time ("the years") somehow just seems to have "flown by." That's why the older we get, the faster time seems to pass ("The holidays are here already! Last Christmas seems like just a few months ago").

Here's the tradeoff: if you want to slow up the subjective sense of time, do little (sit in a lot of waiting rooms). But you'll pay a double price: while waiting, the time will seem endless. However, if you have enough of these experiences, time will seem to have elapsed awfully fast in retrospect.

Differences in people's ability to correctly estimate the passage of time leads to some challenging social implications. I believe physiological differences in time estimation play a role in some people's inexplicable and often maddening inability to be "on time." We all know people who never can seem to arrive at a given place at an agreed-upon time. Their tardiness accounts for many hard feelings, broken friendships, failed romances, and missed career advancements. In most cases we react with anger at such chronic lateness because we resent the other person's apparent disregard for our time and our feelings. But I'm convinced that, rather than an expression of hostility or disrespect, chronic lateness often stems from variations from one person to another in the ability to estimate time. A simple way of testing this hypothesis that I use in my office is to activate a chronometer and ask my patient simply to indicate to me when he thinks that two minutes have elapsed. On occasion I have observed both over- and under-estimation errors of 30 seconds or longer. Fortunately time estimation can be improved. You can also keep a subconscious check on your interval timing system by setting alarms on your watch and trying to anticipate the alarm before it sounds. Since you will be occupying your brain with other matters in the interval, it may take a while to learn to subconsciously synch subjective and objective time.

Another approach to increasing your time sense (as well as your reasoning abilities) is through stories and cartoons. Have someone cut out several comic strips of more than eight or nine frames and see how fast you can assemble them in the order in which events must take place for the comic strip to make any sense. This is similar to the basic skill that we use when we make plans, tell stories, and anticipate what might happen next. People who lack this ability have great difficulty making plans. Along with their difficulty in sequencing, they also often fail to anticipate the likely consequences of their actions. They aren't good at sticking to budgets, organizing a party or other social event. They start too early, or wait until it's too late for other people to make themselves available.

COMIC TIMING *(Time)*

ANSWERS ON PAGES 187–188

▶ *This puzzle exercises your ability to put events in time order.*

In each of the following comic strips, the frames are out of order. List the frames in the order in which they should be to tell a logical story. To solve this puzzle, you will have to understand what the story is and which events must follow which.

1

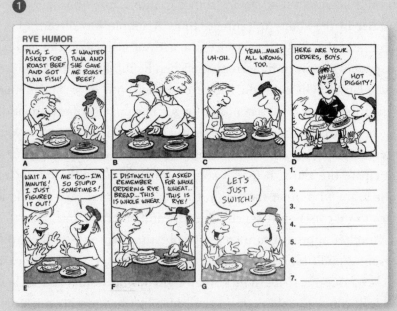

Puzzle and illustration by Robert Leighton

2

Puzzle and illustration by Robert Leighton

3

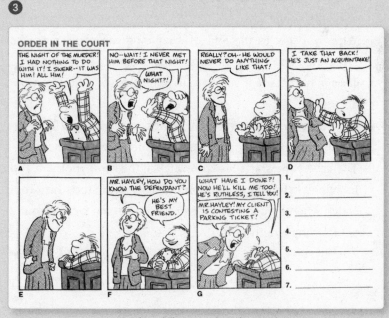

Puzzle and illustration by Robert Leighton

Accurately estimating durations of time is a skill that can be developed. Here is a time estimation game that simulates the experience of singing a song—with no actual singing required.

SILENT SINGING *(Time)*

▶ *This game exercises your ability to imagine time intervals accurately.*

This game tests your ability to imagine the timing of a familiar song. You'll need a stopwatch and a music player.

- Choose a familiar song. Listen to it several times, till you can imagine the entire song from beginning to end.
- Time the song to see exactly how long it is in minutes and seconds.
- Turn off the music player and get your stopwatch ready. Start the stopwatch and imagine listening to the song from beginning to end. Don't look at the stopwatch while you are imagining the song.
- When you get to the end of imagining the song, stop the stopwatch and check the time. How accurate were you?
- Try again and see if you can imagine the song timing more accurately.
- You can play this game with friends. It works best if everyone has his own stopwatch. Everyone listens to the song. Then everyone starts his stopwatch at the same moment and imagines listening to the song. Each player stops his stopwatch when he comes to the end. The person closest to the actual song length wins.

Puzzle and illustration by Robert Leighton

Puzzle and illustration by Robert Leighton

③

Puzzle and illustration by Robert Leighton

REFLECTION. How were you able to put each story in order? What cues told you which panels went after other panels? Which of these three stories was hardest to put in order? Why?

Cognition

9

THINKING IN WORDS: THE HAMMER, THE SAW, AND THE HATCHET

Because of the development of the frontal and temporal areas of our brain, we are primarily word- and concept-creating creatures. And thanks to the introduction of new words and the disappearance of old words through disuse, our language is alive and dynamic. Categories are the natural result of this evolution. We establish our earliest categories by noticing similarities between objects and, as a result, place them in the same category. But our ability to do this varies according to time, place, and circumstance. As children, we experienced great difficulty in performing the acts of abstraction needed to formulate the highest order of categorization. We considered a chair and a table alike, for instance, because we noticed that they both had four legs or were usually made of wood. With additional brain growth and maturation, we came into contact with additional chairs and tables, and as a result we recognized that not all of them have four legs and they aren't always made of wood. Segueing from such observations, we arrived at the insight that both items fit into the more abstract, less concrete category of *furniture*.

A similar continuum from concrete to abstract thinking existed within cultures prior to the early twentieth century. And we're not talking hundreds of years here. In one famous study from 1931 carried out by the Russian psychologist Aleksandr Luria, illiterate peasants in a remote village in Uzbekistan were shown drawings of a hammer, a saw, a hatchet, and a log and asked the following question:

"Which of these things could you call by one word?"

What follows is a typical answer sequence:

A: "How's that? If you call all three of them a 'hammer' that won't be right."

Q: "But one fellow picked three things—the hammer, saw, and hatchet—and said they were alike."

A: "A saw, a hammer, and a hatchet all have to work together. But the log has to be there too."

Q: "Why do you think he picked these three things and not the log?"

A: "Probably he's got a lot of firewood, but if we'll be left without firewood, we won't be able to do anything."

Q: "True, but a hammer, a saw, and a hatchet are all tools."

A: "Yes, but even if we have tools, we still need wood—otherwise, we can't build anything."

This dialogue is an illustration of what intelligence researcher James R. Flynn refers to as seeing the world through "prescientific spectacles."

According to Flynn, when Americans in 1900 were asked, "What do dogs and rabbits have in common?" they were likely to respond, "You use dogs to hunt rabbits." Flynn observes that the correct response today, that both dogs and rabbits are mammals, "assumes that the important thing about the world is to classify it in terms of the categories of science."

During the last three quarters of the twentieth century, according to Flynn, "a vast liberation of the human mind" has been achieved by the adaptation of "the scientific ethos, with its vocabulary taxonomies, and detachment of logic and the hypothetical from concrete referents . . . More formal schooling and the nature of our leisure activities have altered the balance between the abstract and the concrete."

Less sophisticated thinking patterns not only overemphasize concrete relationships but also include metonymy: the substitution of one word for another word that is associated with it in some way. In most instances a part or component is used to represent the whole. "With the bases loaded the coach brought in from the bullpen Big Lefty, who then proceeded to end the inning and save the game by striking out the next three batters." Here "Big Lefty" is shorthand, a metonymy, for a specific left-handed pitcher. "She bought herself a Picasso" is another example in which everyone intuitively recognizes that the buyer purchased one of many possible artworks by Picasso. A metaphor or an analogy based on metaphor is far subtler than a metonym. While primitive prescientific thinkers had no problem understanding metonymy, they could not process metaphor. Such phrases as "the teeth of a comb" or "the face of a crystal" or "the face of a cliff" would create the same befuddlement as Luria's questions. (Such overreliance on concrete relationships can still be found among young children and people afflicted with certain types of brain damage or mental illness, notably schizophrenia.)

Puzzles and exercises can strengthen the ability to replace functional relationships (dogs chase rabbits) and metonymy (parts taken to represent the whole) in favor of abstract similarities (dogs and rabbits are both animals, mammals, etc.). One way of strengthening this power is solving analogram puzzles such as the one on the following page.

B

ANALOGRAMS (Thinking in Words)

ANSWERS ON PAGE 201

▶ *This puzzle challenges you to think about meanings of words and how they are related.*

Complete each analogram by choosing two words from the list below. For instance, ARM is to HAND as LEG is to FOOT, because the FOOT is attached to the end of the LEG in the same way that the HAND is attached to the end of the ARM. Each word is used only once.

1 ARM is to HAND as _____ is to _____.
2 APPLE is to JUICE as _____ is to _____.
3 MOUNTAIN is to PEAK as _____ is to _____.
4 LOAF is to BREAD as _____ is to _____.
5 EGG is to SHELL as _____ is to _____.
6 MOUSE is to CHEESE as _____ is to _____.
7 XYLOPHONE is to STICK as _____ is to _____.
8 LIGHTNING is to ELECTRICITY as _____ is to

_____.
9 BOWLING is to PINS as _____ is to _____.
10 SQUARE is to DIAMOND as _____ is to _____.
11 DIAMOND is to GRAPHITE as _____ is to _____.
12 PUZZLE is to BRAIN as _____ is to _____.

ARCHERY	HEART	RUNNING
BOW	HOUSE	SHEET
BREAD	ICE	SWEATER
CHEDDAR	LEG	TARGET
CHEESE	PAPER	TIMES
CRUST	PEANUT	VIOLIN
ELEPHANT	PLUS	WATER
FOOT	ROOF	YARN

Analograms sharpen your understanding of the dictionary definitions of words. But in real life, dictionary correctness is not what matters. Listen carefully to a conversation between two friends who know each other well, and you will hear all sorts of half sentences, words out of order, and peculiar references known only to the speakers. Anything that triggers the right thought in the other person is fair game.

Here is a game that explores the social nature of communicating in words.

CATEGORIES *(Thinking in Words)*

▶ *This game exercises your working memory.*

This is a party game for four or more people. Each person needs paper and a pencil.

One player, the leader, calls out a category. It can be mundane, such as "animals," obscure, such as "musical instruments from the Middle Ages," imaginary, such as "cities on the moon," or even nonsensical, such as "colorless green ideas."

Each of the other players privately writes a list of five items belonging to that category. The items do not actually have to be correct, since the goal is not to be correct but to guess what other people will write down.

When all players are finished writing, each player reads the items in his list out loud to the rest of the group. Whenever another player hears an item that is also on his list, he shouts out "Match!" and all players who wrote down that item score a point.

The player with the most points wins. The game is then repeated with a new leader naming the category.

This game is quite similar to the commercially available board game Scattergories except that it allows much wilder categories. As with most party games, winning is not so much the goal as having fun and psyching out the other players.

Y

Some other games that exercise your verbal skills include crossword puzzles (great for expanding your vocabulary), Apples to Apples (choose which of several words best fits with a given word), and Taboo (define a given word without using any of five forbidden words).

Crosswords are the most popular type of puzzle in the world. Many people do crossword puzzles regularly to keep their brains sharp. Crosswords exercise a wide range of verbal skills: recalling knowledge of the world from written clues, making associations based on puns and wordplay, and finding words that fit certain patterns of letters. The following crossword puzzle was constructed for this book by Jeremiah Farrell. The theme is "Games."

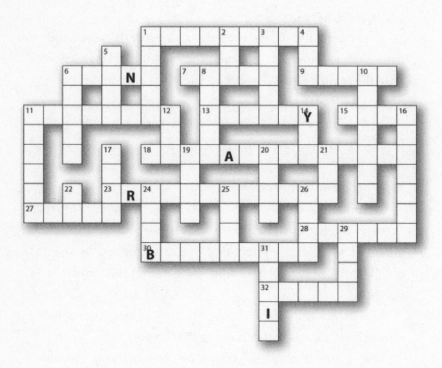

ACROSS

1 Game participant

6 Sports palace

7 Chopped up (and not, as it may sound like, a game involving chance)

9 Swap, as in baseball cards or players

11 Needed for many modern games

13 Not out of bounds, for a ball (2 words)

15 Throw in or hit over 4 down

18 "Come, Watson, come! _____" (And he doesn't mean a soccer ball) From Doyle, *The Adventure of the Abbey Grange* (4 words)

23 English checkers grid (not to be confused with an army recruiting station)

27 Game piece for 22 down

28 Something thrown in the Olympic Games

30 Certain ball game

32 Typical checker color

DOWN

1 Article of sportswear

2 First in a famed pencil-and-paper game

3 Olympic Games award

4 Needed in many games, especially those that use a 15 across

5 Game list on 11 across

6 "In the long run men hit only what they _____"—Henry David Thoreau (2 words)

8 A misplay in hockey

10 Dungeons & _____

11 Musical _____ (popular group game)

12 Cheer heard at football games

14 One response in a game of Twenty Questions

16 Scrabble ingredients

17 Children's game with 21 and 29 down

19 Objects in a spring hunt

20 Was archery this god's game?

21 See 16 down

22 Classic Japanese game

24 Horse common in the sport of kings

25 Setting of TV game show *Cash Cab*

26 Midwest baseball team

29 See 16 down

31 Bridge action

N

GAMES ON THE BRAIN

(Thinking in Words)

ANSWERS ON PAGE 202

▶ *This game exercises your vocabulary.*

Here are a few tips for solving crossword puzzles.

- You don't have to solve clues in order. Skip around, and write in words that you know, then go back and fill in other words.
- If you think you know a word, write it in. You can always go back and erase it if it isn't right.
- Once you have written in a few words, try solving words that cross these words.
- Remember that the theme of this puzzle is "The Playful Brain": many of the words relate to ideas in this book, in a playful way. If you get stuck, look through the chapter titles in the table of contents for inspiration.
- If you get stuck, give the puzzle a rest and return to it later. Sometimes an answer will come to you when you are not thinking about the puzzle.

The classic game Twenty Questions provides a nifty means for testing verbal reasoning, logic, memory, and creativity. You can play it with three other people—as was done on the original Mutual Broadcasting System radio series that first aired on February 2, 1946, from the Longacre Theatre in New York City—or you can use a readily available handheld version. You can also play a solitary game against a devilishly clever neural network–based artificial intelligence system by logging on to www.20Q.net.

If you play the original radio-style game for four people, you can take the role of the answerer, who comes up with a word or subject that the other three players must discover by questioning, or you can be one of the questioners, who must guess the answer in twenty questions or fewer.

If you choose the computerized version, you always play the role of the answerer. In this version, you decide on a word or subject and then 20Q asks the initial question: "Is it animal, vegetable, or mineral?" and things proceed from there. In contrast to the four-player version, the computer variant allows for additional answers (*Maybe, Sometimes, Don't know*, and *Irrelevant*). This is an improvement on the original game, since in some instances answers will vary according to circumstances. For example, if the subject is Panda Bear, the question "Can it be held in the hands?" could be answered as either yes or no, depending on the age of the animal. In this example *Sometimes* would be the best answer. If you play with a new toy computer version, you get to ask the questions.

Twenty Questions can be even more intellectually challenging when the chosen subject is limited to a specific topic, such as Charles Dickens or Star Wars. Obviously such highly specialized variants are possible only when all of the players are knowledgeable about the chosen topic.

But if you want to put your brain to a real test, try an even more interesting version of Twenty Questions devised by the late John Archibald Wheeler, originator of the term "black hole" and one of the most distinguished physicists of the twentieth and early-twenty-first centuries. Wheeler was fascinated with the philosophical implications of quantum physics, especially the finding that electrons or photons can act as either waves or particles, depending on how they are experimentally observed. In order to make this concept clear Wheeler devised a "surprise" version of Twenty Questions. Here is how it's played:

One of the players leaves the room, ostensibly in order for the other players to confer and select a person, place, or thing. When the temporarily excluded player returns, she then begins asking questions of the other players in order to discover what word or subject was decided on by the other players during her absence. But, unbeknownst to the questioner, the other players never made any choice. Instead, a choice will emerge in response to the questions she asks. The process starts when the first person questioned arbitrarily thinks of an object *after* the questioner asks her question. Each of the other persons questioned will do the same, taking care that his response is consistent with all of the previous responses. This requirement places a greater burden on the memories of each of the players than in the traditional Twenty Questions, since the answer the questioner is seeking doesn't really exist but must be created on the basis of the questions as the game progresses.

"The word wasn't in the room when I came in, even though I thought it was," as Wheeler explained it. But gradually, over the space of a varying number of questions, an object finally emerges. Instead of being selected ahead of time in the classic form of the game, the object to be discovered is created during the process of questioning. Try the game for yourself with a group of friends. It's important that everyone pay close attention so that all of the answers are consistent. If that requirement is fulfilled, each answer gradually narrows the choice to a single object—an object that no one selected ahead of time and that could not have been predicted.

1. ARM is to HAND as **LEG** is to **FOOT**.
2. APPLE is to JUICE as **CHEDDAR** is to **CHEESE**.
3. MOUNTAIN is to PEAK as **HOUSE** is to **ROOF**.
4. LOAF is to BREAD as **SHEET** is to **PAPER**.
5. EGG is to SHELL as **BREAD** is to **CRUST**.
6. MOUSE is to CHEESE as **ELEPHANT** is to **PEANUT**.
7. XYLOPHONE is to STICK as **VIOLIN** is to **BOW**.
8. LIGHTNING is to ELECTRICITY as **SWEATER** is to **YARN**.
9. BOWLING is to PINS as **ARCHERY** is to **TARGET**.
10. SQUARE is to DIAMOND as **PLUS** is to **TIMES**.
11. DIAMOND is to GRAPHITE as **ICE** is to **WATER**.
12. PUZZLE is to BRAIN as **RUNNING** is to **HEART**.

REFLECTION. Which analogies did you complete first? Which did you save for last? Which were hardest to solve? Why?

Thinking in Words: The Hammer, the Saw, and the Hatchet

REFLECTION. Which clues did you solve first? Which were harder to solve? What insights helped you solve the harder clues? What did you do when you got stuck?

10

LOGIC: REASONING IN
UNCERTAIN SITUATIONS

During sales the neighborhood electronics store is filled with eager customers. There is no sale today. Therefore the store doesn't have many customers. Does this conclusion logically follow from the premise?

Traditionally, logicians have suggested deciding about such syllogisms by substituting letters for specific content terms in order to avoid being confused by one's previous experiences with electronics stores, sales events, and customer volume. The syllogism in this instance becomes: *If P (a sale), then Q (many customers). Not P (no sale), therefore no Q (store not crowded).* Our brain has greater difficulty processing this form of logical reasoning ("If not . . . then not . . .") as opposed to affirmative "If . . . then" statements (*If the store is crowded, then a sale must be going on,* or *The store is crowded, therefore a sale is going on*). In this instance, the conclusion seamlessly emerges from the premise that indicates that the store is crowded *only* when a sale is taking place. Despite the greater difficulty of the *If not, then not* syllogism, you were probably able to correctly figure out that the conclusion (the store doesn't have many customers) does not logically

R

follow from its premise, since nothing is stated about usual customer volume: the store may be crowded on any given day, sale or no sale.

But reaching a correct conclusion isn't as easy when you are dealing with abstract concepts. Imagine you see the front sides of four cards, showing an E, R, 2, and 7, respectively. You know that every card has a letter on one side and a number on the other. Now consider this rule: If there is a vowel on one side of a card, then there is an odd number on the other side. What is the minimum number of cards you would need to turn over to determine the truth or falsity of this statement?

Almost everyone will correctly select card E (if an even number is found on the opposite side, the rule is false). Selecting the other required card that proves the rule doesn't prove as easy. That isn't card R, since neither an odd nor an even number will help you decide on the truth of the rule, which says nothing about what may be on the other side of a consonant. Perhaps cards with consonants, too, have odd numbers on their opposite sides, which is a separate issue from what is claimed by the rule: A vowel must always show an odd number on the opposite side of the card. The 7 card is also irrelevant: if there is a vowel on the other side, this is consistent with the rule but doesn't definitely prove it. Thus, if a consonant appears on the other side of the 7, then we haven't learned anything that proves or disproves the rule, since consonants may be on the other sides of both odd and even cards. The 2 card is the correct choice, because a vowel on the other side of the 2 proves the rule false. Although the selection of 2 seems obvious in retrospect, less than 25 percent of people taking the test select 2.

As the developers of the above exercise (the Wason card-sorting task) point out, a vast improvement in performance occurs (90 percent accuracy compared with less than 25 percent accuracy) by replacing the abstract context of numbers and letters with something everyone is familiar with:

soda wine over 21 under 21

Look at the four cards here, on which are pictured, respectively, a glass of soda, a goblet of wine, a middle-aged man, and a young boy. Here is the rule: Anyone drinking wine must be over twenty-one years of age. How many cards would have to be turned over to prove that? Obviously the wine card (anyone pictured on the other side under twenty-one disproves the rule), but what other card? Ninety percent of people tested immediately select the card showing the young boy. Wine on the opposite side of that card nullifies the rule. Turning the card depicting the man and finding wine or soda doesn't prove or disprove the rule. Nor is it helpful to turn over the soda card (adults drink both soda and wine). Only the card depicting the boy effectively proves the rule: it must show soda and not wine. In this example a real-world situation simplifies reaching the solution enormously. Why?

Two Different Ways the Brain Processes Logic

Thanks to our brain's organization, we process information about meaningful material differently from how we process information about abstractions. Put another way, our method of reasoning differs according to what we're reasoning about. Take the preceding examples. While all of us are familiar in one way or another with alcohol consumption and age requirements, electronics stores and sales events, few of us spend much time in situations in which we're required to correlate numbers and letters. This distinction between meaningful material and abstractions is mirrored in our brain's activation patterns.

A pathway running in both directions from the frontal cortex to the parietal cortex activates whenever we reason about arbitrary materials (No A are B). In contrast, realistic, personalized, or emotional statements activate a frontotemporal pathway. This suggests a brain-based explanation for something we encounter every day: intuition sometimes plays a more important role than logic when we decide about personal matters. This isn't simply an example of "emotion" overcoming logic (the usual explanation) but, rather, the activation of the frontotemporal pathway at times when the frontoparietal pathways governing abstract reasoning may be more appropriate.

For example, except in life-threatening emergencies, physicians probably should not treat close family relatives. This is because medicine often involves rapidly combining observations of a patient with the doctor's medical knowledge and clinical experience to form syllogisms such as *A sudden catastrophic headache followed by collapse indicates a brain hemorrhage until proven otherwise.* If a patient experiences such a turn of events, the doctor will logically work out that syllogism (*Sudden severe headaches followed immediately by unconsciousness are potentially life-threatening. This is such a headache. Therefore this is a potentially life-threatening situation*).

But if the patient is a relative, the reasoning process may be affected in ways that could lead to grievous consequences (*Maybe that headache is just a bad migraine, so I'll wait a while before considering the hospital*). In this case, emotionally loaded considerations (a son potentially dying of a brain hemorrhage) activates the frontotemporal pathway. What's really needed is the activation of the frontoparietal pathway, resulting in a cool, dispassionate application of that abstract syllogism linking sudden headaches with brain hemorrhages.

Less emotionally loaded reasoning challenges than life-or-death decisions face us every day, such as: *Who is the best candidate to vote for? Are my child's poor grades the result of bad teachers, as he claims, or is he not studying hard enough?* The conclusions one reaches in response to such questions depend very much on the parts of our brain that we activate. In addition to the frontotemporal and frontoparietal distinction mentioned above, the two prefrontal cortices—those areas farthest to the front of the brain—also show different tendencies to respond according to the completeness of available information.

SCOTT The existence of these two very different pathways reminds me of a story about the brilliant twentieth-century physicist Richard Feynman. Whenever someone told him about a new physics theory, he always imagined a particular situation and applied everything he heard about the theory to that situation. Not only did this improve his understanding of the theory, but also it sometimes allowed him to immediately point out a flaw in the theory by citing a specific situation in which something went awry. By grounding an abstract theory in a concrete example, Feynman was deliberately using his frontotemporal pathway.

I imagine that something similar goes on in skilled mathematicians. The hallmark of good mathematicians is that they have strong intuitions about how to solve a problem, even before they know the answer. If you monitored the brain activity of two people, one skilled in math and the other not, trying to solve the same abstract math problem, I bet you would see the skilled mathematician engage the frontotemporal pathway and thus emotion, whereas the unskilled person would engage only the frontoparietal pathway, where the problem remains unemotional.

I am also reminded of a piece of advice I once heard given to a first-time magazine writer: Imagine that you are writing a letter to a friend, not to an anonymous reader. This, too, engages emotional thinking by making a somewhat abstract assignment more concrete.

Reasoning in Uncertain Situations

The right prefrontal cortex has a critical role to play in reasoning about incompletely specified situations that do not allow for a conclusion (*Kelly is smarter than Donna; Kelly is smarter than Wendy; Wendy is smarter than Donna*). In contrast, valid unambiguous statements (*All cats are mammals; my pet is a cat; my pet is a mammal*) engage the left prefrontal cortex.

Problems arise when dealing with seemingly valid statements that can't be proven. For instance, the choice of one candidate over another in an election is indeterminate, i.e., nobody knows for certain how well a given candidate will perform if elected. So why do committed Republicans and Democrats often claim that their candidate is the only "reasonable" choice and thereby express a degree of conviction that would be valid only for an unambiguous logically valid proposition?

According to Michael Gazzaniga, director of the SAGE Center for the Study of the Mind at the University of California, Santa Barbara, the left hemisphere of the brain operates as an "interpreter" that abhors uncertainty and automatically comes up with explanations that may or may not be true. "Any time our left brain is confronted with information that does not jibe with our knowledge or conceptual framework, our left hemisphere interpreter creates a belief to enable all incoming information to make sense and mesh with our ongoing idea," writes Gazzaniga in his book *The Ethical Brain: The Science of Our Moral Dilemma.*

In the political example, the left-hemisphere interpreter picks up on and incorporates only those facts about a favorite candidate that support the preconceptions of the ideologically driven voter. Think of election results as dependent on how many people's interpreters create reasons to vote for a candidate. In Gazzaniga's words, "The left hemisphere 'interpreter' constructs theories to assimilate perceived information into a comprehensible whole. In doing so, however, the elaborative processing has a deleterious effect on the accuracy of perception."

In short, reasoning is best considered by referring to what neuroscientist Vinod Goel of York University in Toronto calls "dual neural pathways": a frontotemporal pathway for processing familiar material, and a frontoparietal pathway for processing unfamiliar, abstract material such as letters and numbers. And in those situations where it's difficult to decide about the correctness of a proposition because of ambiguity, it's necessary to take full advantage of the right hemisphere's role in preventing overinterpretation by the left hemisphere's interpreter.

SCOTT I also do something like that, so as not to be stymied when confronted with mathematical puzzles, for instance tricks I can use for getting a handle on 3-D rotation of objects, or tricks for not getting stumped by pure logic. In my case, the tricks involve considering particular cases (like Feynman) but also validating things several different ways, and having overall intuition about noticing when a situation is slippery to think about, and taking appropriate precautions to represent the problem properly.

Following are some puzzles that challenge your reasoning abilities using both practical and abstract examples. Notice as you work them out how much harder it is to correctly solve puzzles presented in abstract rather than familiar contexts. One of the reasons for this discrepancy is that the "interpreter" plays less of a role in puzzles involving letters, numbers, and other abstract items than it does in puzzles involving people and everyday situations.

VISUAL LOGIC *(Logic, Visual Thinking)*

ANSWERS ON PAGES 218–225

▶ *This puzzle exercises your ability to think through a problem logically with the aid of pictures.*

Here are three logic puzzles that are easier to solve if you draw a diagram. See if you can discover what type of diagram works best for each puzzle.

① APPLES AND ORANGES

Three closed boxes contain apples and oranges in various combinations. One box is labeled *Apples*, the second is labeled *Oranges*, and the third *Apples & Oranges*. You know that all three boxes are mislabeled. You may ask a friend to remove one piece of fruit from one of the boxes and show it to you. On the basis of that information, you must deduce how to place all the labels correctly. Which box should you ask your friend to open, and how will you deduce the correct label placements?

▶ ▶ ▶

② WHERE DID FAY SIT?

Abe, Barb, Carl, Deb, Ed, and Fay are seated around the circular table shown above. Use the following clues to deduce in which chair Fay sat.

1. Deb sat directly opposite Abe.
2. Ed sat just to Carl's right.
3. Abe, Barb, and Carl sat with no two of them next to each other.
4. Barb sat in seat 1.

③ MOVIE MENU

Put the following movies in order so the first letter of each movie title appears in the next movie title.

BRAZIL
CASABLANCA
CHARADE
DETOUR
GOODFELLAS
IKIRU
METROPOLIS
PSYCHO
UGETSU

L

Abstract Reasoning

Purely abstract reasoning of the "P and not Q" sort is difficult for the mind to grasp. Only with discipline can we force our minds to tread the narrow paths of logical reasoning. But with difficulty comes challenge. Just as adventurers thrill to the challenge of reaching the top of Mount Everest or plumbing the depths of the Marianas Trench, so puzzle lovers thrill to the challenge of clambering up the slippery slopes of logic puzzles like Sudoku.

Sudoku began modestly in America in 1979, moved in 1986 to Japan, where it was polished and refined, and gained international fame when it premiered in *The Times* in Britain in 2005. Sudoku is now popular worldwide in both printed and computer forms, and figures prominently in Nintendo's trailblazing electronic brain exercise game *Brain Age*.

Although Sudoku uses numbers, it is really a game of pure logic—no calculation is required. You never have to guess. Every number can be deduced through careful reasoning. Although the rules of Sudoku are simple, the reasoning can get surprisingly deep. Here is a set of puzzles that take you step by step through some of the key Sudoku strategies, starting from the simplest, and getting gradually harder.

SUDOKU *(Logic)*

ANSWERS ON PAGES 225–228

▷ *This game exercises your logical-thinking skills.*

The challenges that follow, while not conventional Sudoku, give you the opportunity to hone your chops. The goal in each puzzle is to complete the grid so that every row, column, and outlined region contains the numbers 1 through 5. (In conventional

▶ ▶ ▶

Sudoku the board is larger, and uses the numbers 1 through 9.) Every puzzle has a unique answer.

The simplest strategy for attacking Sudoku puzzles is to look for a row, column, or region that has only one empty square. Because all numbers in a row, column, or region must be different, the last empty square will always contain the one unused number. Use this technique to solve the first three puzzles.

❶

4	5		1	2
1	3		2	4
3	4		5	1
2	1		3	5

❷

	2	1	5	3
2		5	4	
5	1			4
1		2		5
	5	4	1	

▶ ▶ ▶

❸

1	2	4		
2		1		4
3			2	1
4		3	1	
	1			3

There are various strategies for deducing where a number should go in a Sudoku puzzle. A few of these are illustrated below. Can you explain why the starred square in each of these partially completed grids must contain the number 5? Note: These partial boards do not contain enough numbers for you to figure out a complete unique solution.

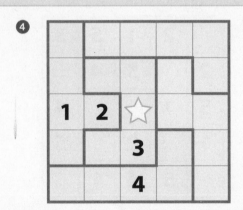

❹

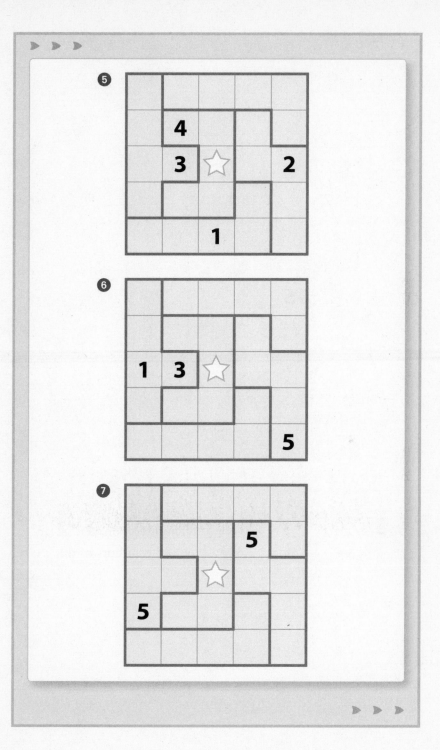

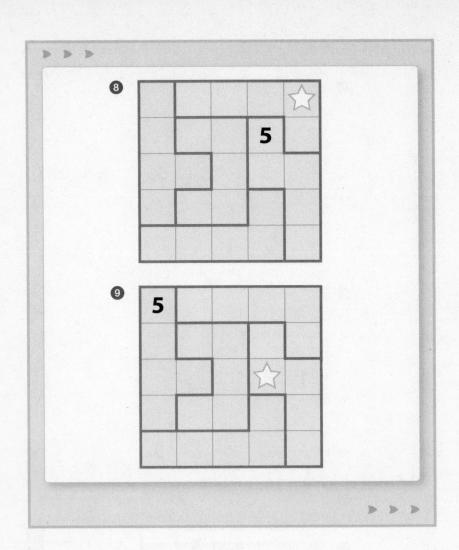

▶ ▶ ▶

Now use your newly acquired Sudoku strategies to solve these difficult five-by-five puzzles.

⑩

	3			
4		5		
	2	3		

⑪

4			3	
		5		
				1

⑫

				5
			4	
		1		
	2			

ANSWERS TO VISUAL LOGIC

1 The answer is that you should ask your friend to take one piece of fruit out of the box labeled *Apples & Oranges*. The reasoning to solve this puzzle is simple but confusing. To help you understand the answer, I've drawn some diagrams. Here is how I solved this puzzle.

Ask your friend to remove one piece of fruit from the box labeled *Apples & Oranges*. Let's say your friend removed an apple.

Since you know all three boxes are mislabeled, this box must contain only one kind of fruit, in this case apples. Let's record this fact:

Now turn to the box labeled *Oranges*. It is mislabeled, so it can't contain only oranges, and you have just deduced that another box is the one that contains only apples. Therefore it must contain apples and oranges.

Finally, the remaining box, labeled *Apples*, must contain the remaining possibility—oranges. And that completes the solution. Similar reasoning works if the first piece of fruit is an orange instead of an apple.

2 Fay sat in seat 2. This puzzle is harder to solve than the Apples and Oranges puzzle, but again, diagrams help a great deal. Here is how I solved this puzzle.

Clue 3 says that Abe, Barb, and Carl sat with no two of them sitting next to each other, so they must have sat at the corners of a big triangle, with Deb, Ed, and Fay between them. Something like this:

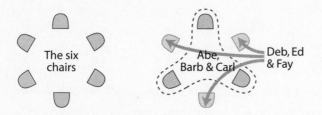

Clue 4 says that Barb sat in chair 1, then Abe and Carl sat in seats at 3 and 5, although we don't yet know which of the two sat in which chair. Let's consider each possibility in turn.

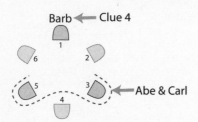

Let's assume Abe sat in chair 5. Then we know that Carl sat in chair 3, at the third corner of the Abe-Barb-Carl triangle.

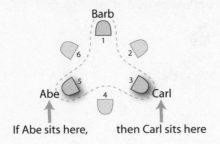

If Abe sits here,   then Carl sits here

Clue 1 says that Deb sat opposite Abe, which would be chair 2.

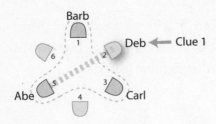

But clue 2 says that Ed sat just to Carl's right, which would also be chair 2. We can't have two people in the same chair, so our assumption that Abe sat in chair 5 must be wrong.

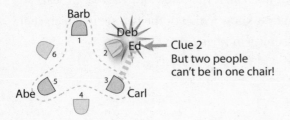

Therefore Abe sat in chair 3 and Carl sat in chair 5 . . .

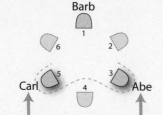

Therefore Carl and Abe sat in these chairs

. . . from which we can deduce that Deb sat in chair 6 (opposite Abe), and Ed sat in chair 4 (just to the right of Carl).

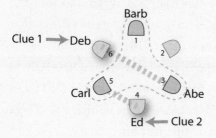

That leaves chair 2 empty, which is where Fay must have sat.

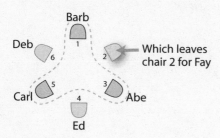

● Here is the unique solution:

DETOUR

GOODFELLAS

UGETSU

IKIRU

BRAZIL

CASABLANCA

CHARADE

PSYCHO

METROPOLIS

This puzzle is very difficult to solve without a diagram. To find the solution, arrange all the movie titles in a circle. Draw arrows that connect each movie title with every other title that could follow it, as shown below. For instance, draw an arrow from DETOUR to CHARADE because CHARADE can legally come right after DETOUR (the first letter of DETOUR is in CHARADE). The solution will be a chain of arrows that visits every movie title just once.

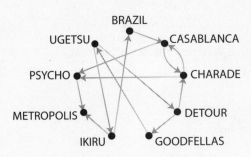

This diagram helps us see patterns that are hard to see if you look just at the list of titles. For instance, we can easily see that there are no arrows leaving METROPOLIS (no other title contains the letter M), so METROPOLIS must be the last title in the list.

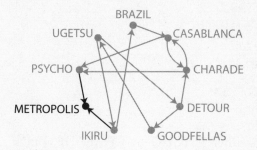

There are two arrows leading into METROPOLIS, which means it can be preceded by either PSYCHO or IKIRU. Which one is correct? Note that there is only one arrow leaving PSYCHO, while there are two arrows leaving IKIRU. We know that PSYCHO isn't the last movie, so it must precede METROPOLIS. So the last two movies must be PSYCHO and METROPOLIS.

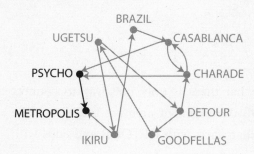

Each of the movies IKIRU, BRAZIL, and GOODFELLAS has only a single arrow leaving it, so these arrows must be part of the answer.

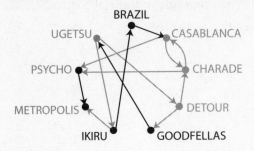

CASABLANCA can lead to either PSYCHO or CHARADE. But if CASABLANCA leads straight to PSYCHO, then CHARADE has nothing that can follow it. Therefore CASABLANCA must lead to CHARADE, which must lead to PSYCHO.

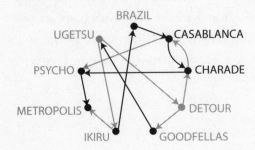

Finally, we can quickly see that there is only one way to connect UGETSU and DETOUR into our chain of arrows. And there's the solution: a chain of arrows that starts with DETOUR and ends with METROPOLIS.

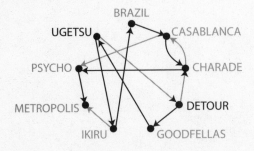

REFLECTION. Which of these three logic puzzles were you able to solve? How did you solve it? Did you draw diagrams? Did you use a technique similar to the one I used, or did you do it a different way?

ANSWERS TO SUDOKU

1

4	5	3	1	2
1	3	5	2	4
5	2	1	4	3
3	4	2	5	1
2	1	4	3	5

2

4	2	1	5	3
2	3	5	4	1
5	1	3	2	4
1	4	2	3	5
3	5	4	1	2

3

1	2	4	3	5
2	3	1	5	4
3	4	5	2	1
4	5	3	1	2
5	1	2	4	3

4

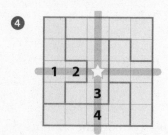

The numbers 1, 2, 3, and 4 occur in the same row or column as the starred square. That square must therefore contain the number 5.

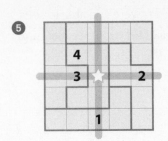

The numbers 2 and 3 occur in the same row as the starred square. The number 1 occurs in the same column, and the number 4 occurs in the same C-shaped region as the starred square. Therefore the starred square must contain the number 5.

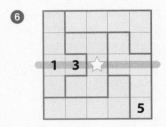

The middle row must contain the number 5. The first two squares of the row are already occupied, and the last two squares are in a region that already contains a 5. Therefore the 5 must be in the middle square of the middle row.

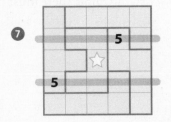

The central backward-C-shaped region must contain the number 5, but the 5 can't be in the top two or bottom two squares of the region, because they lie in rows that already have 5s. Therefore the middle square must contain the 5.

8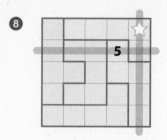

The rightmost column must contain the number 5. Of the five squares in that column, the second square cannot contain a 5 because there is already a 5 in that row, and the bottom three squares cannot contain a 5 because they are in the same region as a 5. Therefore, the top square of the rightmost column must contain a 5.

9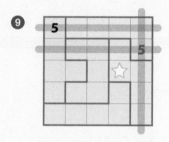

Because there is a 5 in the top row, there cannot be a 5 in the four top squares of the upper right region. Therefore, a 5 (shown in gray) must appear in the one remaining upper right square. Similarly, because this 5 "cancels" four of the five squares in the lower right region, there must be a 5 in the starred square, the one remaining square in the lower right region.

F

⑩

2	3	4	5	1
4	1	5	2	3
1	2	3	4	5
5	4	1	3	2
3	5	2	1	4

⑪

4	1	2	3	5
2	5	3	1	4
5	3	1	4	2
1	4	5	2	3
3	2	4	5	1

⑫

1	3	4	2	5
5	1	3	4	2
2	5	1	3	4
4	2	5	1	3
3	4	2	5	1

11

EMOTIONS AND THINKING:
THE ANGER SUPERIORITY EFFECT

So FAR we've overlooked a most important contributor to logical thinking (as well as memory and physical and mental performance in general): *emotions*. Certainly emotions play a huge role in physical performance. To appreciate this, play a few solitary rounds at darts. Note your performance. Now position a picture of one of your children or a child known to you on the bull's-eye of the board and start again. In the original study testing this rather macabre situation, most people were hard-pressed to hit the child's photo—especially if it was of their own child—despite their recognition that it was only a picture. In other words, full intellectual recognition of the distinction between reality and image wasn't sufficient to offset the emotional reluctance to throw a sharp instrument at an image of their child. Most people are similarly reluctant to tear up a piece of paper with a loved one's name written on it.

Emotions can also influence our mental performance, as with the emotional Stroop test, in which the subject is given a string of words to learn. Emotionally provocative words take longer to learn. The Yerkes-Dodson law (mentioned in chapter 2) illustrates a similar interference effect of emotions upon cognition: anxiety interferes with our ability to sustain the attention required for learning.

Emotions can sometimes influence our thinking in unexpected ways. Deciding which candidate to vote for in an election, for instance, isn't simply a matter of consciously deliberating about the qualifications of the competing candidates. Nor are political views, contrary to popular belief, entirely explained on the basis of background and individual experiences. Equally important are certain physiological traits known to be linked with emotions. Two traits stand out: skin conductance and the blink reflex.

Heightened arousal leads to increased moisture in the outer layers of the skin, thus enhancing electrical conductance. This provides an indirect measure of sympathetic nervous system activation. The second measure of arousal, the blink reflex, increases in amplitude when a person is fearful or anxious. In one intriguing psychological experiment, testing skin conductance and blink reflexes, individuals who overreact to sudden loud noises and threatening visual images (a picture of a spider crawling over the face of a frightened person, an open wound filled with maggots) tend to favor political policies aimed at maintaining the existing social order (defense spending, warrantless searches, capital punishment). People with less reactive skin conductance and eye-blink reflexes, on the other hand, tend to favor more open, more expansive, and less defensive orientations such as foreign aid, immigration, and international compromise.

According to the authors of this study, published in the standard-setting journal *Science*, "The degree to which individuals are physiologically responsive to threat appears to indicate the degree to which they advocate policies that protect the existing social structure . . . Political attitudes vary with physiological traits linked to divergent manners of experiencing and processing environmental threats."

Psychologists have found this powerful effect of emotions on reason to be especially true among people who are undecided and not allied with a particular party or point of view. In a provocative paper, "Automatic Mental Associations Predict Future Choices of Undecided

Decision-Makers," published in *Science,* Professor Silvia Galdi and her associates measured the time required for a response and error rates as their subjects looked at rapidly presented pictures of items related to a controversial political topic (the enlargement of a U.S. military base in the local community). They found that a subject's physiological responses to the pictures (increases in skin conductance and the amplitude of the eye-blink reflex) provided a reliable indicator of what subjects would eventually decide about the military base when asked at a later date—even though when Galdi had tested them a week earlier, they claimed that they had not yet decided how they felt about the base.

Among undecided voters, automatic, spontaneous and rapid responses to pictures of the candidates, quotes from the candidates' speeches, and automatic mental associations to the key campaign issues may be better determinants of how they will vote than anything they might say about their political preferences. Although voters may say one thing about their probable votes, their actual votes are often more accurately predicted by their unconscious automatic associations.

A similar view of the importance of the brain and nervous system in powerfully influencing our political and cultural beliefs was espoused by William James, the most influential and acclaimed psychologist in the nineteenth century. In his essay "The Sentiment of Rationality" he advocated the radical view (for the time) that many of our most deep-seated views on how the world should "work" aren't based on purely rational considerations but result from physiological predispositions ("sentiments," as James called them).

Among the brain structures responsible for this robust influence of emotions on our thinking, the amygdala turns out to be most important. This tiny structure, located deep in each of the temporal lobes, is the most important component in a network of nerve fibers known as the limbic system. While all the components of the limbic system are related to emotional perception and expression, the amygdala is especially important because of its vast network of connections with just about every part of the brain. When you "freeze" upon encountering a snake during your walk in the woods your almost instantaneous response results from an early-warning system provided by your amygdalae, which responded to the snake even before you consciously perceived it. Alerting and orienting to danger—that's the function of the amygdalae. They do this by acting as a conduit linking the cerebral hemispheres, which integrate the sights, sounds, and other perceptions responsible for what we're experiencing from moment to moment; the motor control centers below the cerebral hemispheres; and the other emotion-producing components of the limbic system.

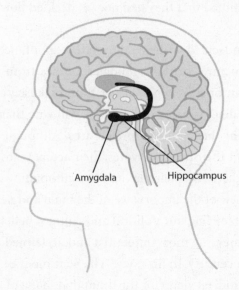

Amygdala   Hippocampus

When you jump back from the curb in order to avoid being struck by a car, it's your amygdalae that first recognize the danger and trigger the muscle control centers to move quickly. It's your cerebral hemispheres that, seconds later, provide you with an explanation of what happened. But there's another component to your experience: a racing heart, a sinking feeling in the pit of your stomach, a sense of breathlessness. Finally, you may feel frightened, angry, and violated in response to your harrowing experience.

Recently, neuroscientists have come up with refinements in our explanation of the mechanisms responsible for our emotions. They started from the long-established premise that emotions result from the activation of what's called the autonomic nervous system (ANS). As the name implies, the responses of this system remain outside of conscious control. Included here are increases or decreases in heart rate, blood flow to the face and neck (blushing from embarrassment or blanching in fear), and sweating, among other responses. Your racing heart, your breathlessness, and that sinking feeling in your stomach resulted from the action of your autonomic nervous system. But everything doesn't occur automatically. Consciousness obviously plays a role in our emotions. When we see a wild bear approaching us, our knowledge that the bear is dangerous (stored in the cerebral hemispheres) seems to be responsible for the train of events set off within the ANS (racing heart, sweating, and so on) that leads to our rapid exit from the scene. I use the qualifying word "seems" because another explanation is also possible. "Do we run from a bear because we are afraid, or are we afraid because we run?" asked psychologist William James. He suggested that the physical changes accompanying the act of running (breathlessness, racing heart, and the like) created the experience of fear.

Neuroscientists now believe that different emotions result from specific patterns of autonomic nervous system activation. For example, if someone, without telling you what he is doing, gives you muscle-by-muscle instructions on how to mimic the facial expressions characteristic of anger, sadness, disgust, happiness, or any other emotion, this modeling will induce changes in your autonomic nervous system (such as alterations in heart rate, skin temperature) that are characteristic for that specific emotion *even though you're not consciously aware of the emotion simulated by your facial expression*. Moreover, the autonomic responses are strongest when your mimed facial expression most closely resembles how your face would appear if you were *really* angry, or sad, for instance. At some point—it varies from one person to another—the act of miming the facial expression of an emotion will induce in you the subjective experience of that emotion. Neuroscientists explain this as the result of brain signals simultaneously activating the muscles responsible for the facial expression of an emotion, as well as the circuits responsible for the subjective experience that accompanies

R

that emotion. Actors are intuitively aware of this process and, with proper training, can both imitate a facial expression and subjectively experience that emotion.

Actors are at one end of a continuum measuring people's ability to *perceive* and *express* emotions. At the other end are people who, when it comes to perceiving other people's emotions, just never seem to "get it." They're completely oblivious to the exasperated tones, rolling eyes, and angry looks that we all use to signal our displeasure. Differences are also encountered when it comes to the expression of emotion. While actors are virtuosos in expressing emotion by means of voice, gesture, and body language, other people (we all know someone like this) are distinctly robotic. At all times they speak in a pure monotone, varying little from one sentence to another. In response to situations that could be expected to evoke anger, joy, sadness, or other emotions, their voices sound flat, completely lacking any of the usual changes in pitch, timbre, and volume that characterize a person experiencing emotion. One such person, a teacher described by neuroscientist Dale Purves, couldn't maintain discipline in her classroom or at home with her own children because her students and her children couldn't tell when she was angry or upset. As a result, she had to convey her inner feelings with such announcements as "I am angry and I really mean it." Nor is such absence of emotion all that uncommon, I believe. Many a marriage founders as a result of the perception by one of the partners that his or her mate (gender doesn't seem a relevant variable here) doesn't "act like he [or she] loves me."

The right hemisphere's emotional dominance also plays out in the visual sphere. Since the right hemisphere controls movements of the left side of the face, emotions can be more accurately and easily identified by looking at the left side of a person's face.

Find a full-face photograph in a magazine. The face should look directly at you, with both sides of the face equally visible. Cover one side of the face, then the other. In most cases (if it is really full-face) you will notice that the left side of the face is more emotionally expressive. In unsmiling photographs, the left side of the face is still more likely than the right to provide an indication of the subject's emotions at the time the photograph was taken or in some cases shows a permanent personality trait. But a word of caution is in order: "reading" facial expressions or tones of voice involves a certain subjectivity that can be overcome only by professional training. So whenever you evaluate another person's emotions on the basis of listening or observing, be prepared to give that person the benefit of the doubt.

The link between emotions and their facial expressions is now sufficiently recognized that many of us freely sprinkle our Internet communication with "emoticons"—you know, those cutesy smiling or frowning faces we all like to put on our IMs and text messages. The word *emoticon* is a mix of *emotion* (or *emote*) and *icon*. While cartoons of the human face are a popular emoticon, many Internet users prefer the earliest emoticons (dating to 1857) formed by text representations. Here are a few of them:

:- for tongue in cheek, with the hyphen representing the tongue and the colon the teeth
-: for sticking out the tongue as an expression of anger or derision
:-) when rotated 90 degrees becomes ☺

In the following puzzles, let's restrict ourselves to emoticons involving the human face, which is processed in the brain in the fusiform gyrus located at the junction of the inferior occipital and temporal lobes. And since we're social critters, the fusiform gyrus operates with hair-trigger effectiveness. For instance, an angry face in a crowd will "pop out" when hidden among friendly or neutral faces. It doesn't matter how many faces are in the display: the *anger superiority effect* (the term used by the scientists who first observed and named it in 1988) is just as powerful for a large array as for a small one. Friendly faces, in contrast, don't pop out and must be looked for via a sequential search.

Why are we so much better at detecting angry/hostile faces than friendly faces? Evolutionary biologists claim that survival mechanisms require quick detection of potential threats in order to avoid death or injury resulting from the aggressive response of the angry person. Normally, we all respond to the emotional expressions on other people's faces—recognizing anger, for instance—even faster than we recognize who they are: we experience fear at the angry face before we know who it is that we're afraid of. The action of the amygdalae, those almond-shaped structures attached to the hippocampus on each side of our brain that process emotions, can be sufficiently powerful that we may become fearful under circumstances where we have nothing objectively to be fearful about, i.e., while watching a horror movie.

Even people with poor facial recognition abilities ("He looks familiar, but I can't quite place him") perform normally in the detection of an angry facial expression. In this instance, memory is intact for emotional expression detection despite the inability to know or recognize who the person is. This comes about thanks to the action of the amygdalae.

While the above explanation for our sensitivity to other people's emotions (especially negative ones) seems to make sense, Mark Baldwin, an expert in social cognition at McGill University in Montreal, has another explanation. He believes that the pop-out effect is more related to low self-esteem:

"People with low self-esteem seem to have a tendency to monitor the environment for rejection rather than acceptance." According to Baldwin, individuals with low self-esteem are predisposed to detect threatening or angry faces. Furthermore, Baldwin believes, this tendency can be reversed by what he calls "bias reduction training," i.e., learning to pick out friendly rather than hostile or angry faces.

Even if your self-esteem is quite in order, you'll find that identifying friendly faces in a sea of hostile, angry faces isn't so easy. Test yourself by going to Baldwin's website, www.selfesteemgames.mcgill.ca, and click on "Games" and then on "EyeSpy."

If you want additional practice enhancing your emoticon skills, try the challenging and hard-to-stop-playing cell phone game Emoticons (carriers T-Mobile and Verizon). In this casual game, you click on a specific emoticon with the cursor, thus putting it "in play." You must then click on strings of similar emoticons, which makes room for new emoticons. The more you clear in one click, the more points you get. When you've filled the score meter, you advance to the next level.

EMOTICONS *(Emotion)*

ANSWERS ON PAGE 238

▶ *This puzzle challenges you to identify emotions in facial caricatures.*

EMOTICONS

Match each of the twelve emoticons below with the feeling or facial gesture it expresses. To read an emoticon, turn the book sideways clockwise and look at each group of punctuation marks as if it were a face.

1. Happy	a. XD
2. Laughing	b. ;)
3. Sad	c. :O
4. Tongue out	d. :(
5. Surprised	e. X(
6. Bored	f. :P
7. Confused	g. :'(
8. Shocked	h. :)
9. Winking	i. :S
10. Angry	j. :/
11. Crying	k. :>
12. In love	l. =:O

A

ANSWERS TO EMOTICONS

1. Happy   = h. **:)**
2. Laughing   = a. **XD**
3. Sad   = d. **:(**
4. Tongue out = f. **:P**
5. Surprised   = c. **:O**
6. Bored   = j. **:/**
7. Confused   = i. **:S**
8. Shocked   = l. **=:O**
9. Winking   = b. **;)**
10. Angry   = e. **X(**
11. Crying   = g. **:'(**
12. In love   = k. **:>**

REFLECTION. Do you already use emoticons? If not, was it easy or hard for you to see these symbols as faces? If you do use emoticons, which were familiar and which were not?

12

MATHEMATICS: DOING THE NUMBERS AT THE CHECKOUT LINE

MANY PEOPLE avoid mathematics or mathematical concepts whenever possible. This is unfortunate, since mathematical literacy is an important component of general intelligence. Puzzles can provide an instructive and entertaining avenue toward overcoming "math phobia." As a lead-in, let's take a moment to explore what goes wrong in mathematical reasoning as a consequence of brain damage.

Look at these three subtraction tallies.

138	923	501
− 74	− 644	− 322
64	321	221

Of these three tallies, only the first one is correct. That correct answer was unwittingly arrived at by the application of an invalid and idiosyncratic rule. Can you identify the procedures being used here by an adult with damage to a specific area of his brain? Hint: In the above examples, think of how very young children often approach subtraction: one take away two is impossible, so the correct sequence must be two take away one.

In these examples taken from cognitive neuropsychologist Brian Butterworth's book *What Counts: How Every Brain Is Hardwired for Math*, Butterworth's patient "Signor Tiziano," who suffered a stroke limited to his left parietal lobe, always subtracts the lesser number from the larger number except when subtracting a two-digit number from a three-digit number. In the two incorrect examples involving three-digit numbers, the smaller of the two numbers in each column was simply subtracted from the larger one. In the correct example involving two digits subtracted from a three-digit number, the leftmost two top digits are treated as a single number. Thus, instead of subtracting 3 from 7, the smaller from the larger, he subtracted 7 from 13, unwittingly coming up with the right answer.

Butterworth refers to such examples as illustrating a "subtraction bug," a specific flaw in mental subtraction. Nor is Signor Tiziano an isolated case. Many other brain-damaged patients have turned up over the years who have pointed to the importance of the parietal lobes in number processing. This is especially true in instances where language plays little role.

For example, subtraction problems elicit a stronger parietal response than addition and multiplication problems. Any thoughts about the origin of this difference? It's probably because addition and multiplication facts are encoded in the brain in the form of memorized tables learned in childhood and, as a result, are more likely to be encoded in the brain's language areas. For instance, the question "How much is 7 x 8?" doesn't stimulate a mathematical process but, rather, a response originating in the language areas and based on knowledge established during childhood when learning the 7 and 8 times tables. In contrast, "How much is 56 divided by 7?" takes a bit longer for most people because they never learned a division table. (While it's true that students in France and a few other countries learn division as well as addition and subtraction tables, this is not the norm in most English-speaking countries.)

As a result of the fact that we learn multiplication tables and commit them to long-term memory, elementary multiplication operations are typically retrieved using the language areas of the brain rather than the parietal lobe. Subtraction and division, on the other hand, require quantitative calculations that must enlist the contribution of the parietal lobe.

Here is the general rule: Nonverbal number processing involves the parietal lobes (specifically an area called the intraparietal sulcus), while verbally mediated number processing (usually based on previously learned tables) takes place in the language areas, especially the angular gyrus. "Writing words and writing numerals, reading words and reading numbers, all involve distinct brain circuits, despite having common input pathways from the eyes and common output pathways to the hands," according to Brian Butterworth.

This breakdown of mathematical processing into a series of separate procedures leads to some interesting and, when first encountered, seemingly bizarre real-life situations:

Imagine meeting a fifty-two-year-old woman who has lost the ability to recognize and name her fingers, distinguish her right side from her left, write a complete sentence, and carry out the simplest of calculations. Actually this tetrad of symptoms—dubbed Gerstmann's syndrome in tribute to Josef Gerstmann, the nineteenth-century neuropsychiatrist who first described it—is fairly commonly encountered by neurologists. It results from damage to the left parietal lobe. Although not everyone agrees on the specific explanatory details, the syndrome suggests an intimate association within the parietal lobes between the representation of fingers and the representation of numbers. This association is most evident in young children and, at the other end of the life spectrum, in people afflicted with Alzheimer's disease and other dementias. Both groups regularly resort to finger counting when doing numbers. Even some perfectly normal adults finger count when asked to do calculations, as I have observed in my clinical practice.

Even among high achievers, the "number sense" varies greatly. People termed dyscalculics suffer from a severe impairment in their ability to perform mental calculations despite a normal IQ. When the

F

dyscalculic attempts to perform mathematical calculations, his brain shows dysfunction in the parietal lobe. People who experience no difficulty with counting or other mathematical processing show no such dysfunction.

While the parietal lobes are key to number manipulation and, when damaged, can cause dyscalculia and other serious deficiencies in counting abilities, calculation isn't a single operation but can involve two quite distinct processes. For example, imagine that I have placed 53 marbles on a table in front of you and asked you to count them as quickly as you can without touching them. There are two ways of approaching this challenge. You can count the marbles one by one or you can try estimating the number. In actual tests of people's ability to estimate the correct number of items in such situations, the answers range from 40 to 60 marbles. Estimates such as 10 or 100 are unusual. People immediately recognize that more than 10 marbles are present and 100 marbles seems too high an estimate.

Imagine now that I move 5 marbles to the side and ask you to estimate that number. You will immediately come up with the correct number without having to resort to counting. The time required for the two operations differs significantly. While counting the 53 marbles takes about 250 milliseconds per marble, a quick estimation of the 5 marbles is nearly instantaneous (40 to 100 milliseconds). Why this difference?

When counting 5 marbles, your brain performs a calculating process known as subitizing, a word based on the Latin *subitus* (sudden). Subitizing involving 5 or fewer items is rapid, accurate, and associated with a high degree of confidence: you simply "know" in an instant that you are looking at 5 marbles. However, the larger number of marbles takes longer because you have to resort to actually counting the marbles. When finished, you may want to recheck your calculation because you feel uncertain of your result. Perhaps, despite your best efforts, you even make a mistake. As an additional source of error, the more marbles that are present on the table, the longer it takes to count them. This timing difference even follows a mathematical rule based on response-time measurements: 250 to 350 milliseconds are needed to count each additional marble beyond 4. This lag is sufficiently reproducible to the point that an outside observer can estimate the number of marbles you counted by timing the process and dividing by 300 milliseconds.

Here's another, more commonplace example: When you approach a series of checkout lines in your local supermarket, you have three methods of selecting the shortest line. If there are shorter lines, you can instantaneously estimate by subitizing any line consisting of fewer than five people. (This takes less than a second, and you're consistently accurate.) If the lines are longer, you have two choices: you can count the number of people in each line (a time-consuming process that, in a busy supermarket, has to be updated every few seconds as additional customers enter the lines); or you can simply take a quick look at all of the lines, come up with a rough estimate of the number of people in each line, and then choose the one that seems to be the shortest. All of these methods may produce the same answer, but only one of them involves counting.

An fMRI taken at the time you made your choice in the supermarket would show different activation patterns, depending on how you selected which line to enter. Subitizing activates sites in the visual cortex on both sides of your brain, which is confirmation that the process involves only the early stages of visual processing. Thought was not required: you immediately and automatically and without any sense of effort estimated the number of people (five or fewer).

With lines containing more than five people, counting the number of people in each line activated the visual cortex on each side of the brain, but both parietal lobes also lit up, along with brain areas involved in the maintenance and shift of attention (parts of the frontal and cingulate areas).

Finally, intuitive estimation of the number of people in the various lines, which is essentially guessing, is not associated with activity in the parietal lobes. Only when you deliberately carry out the process of counting do the parietal lobes come into play.

In summary, different brain areas are activated when you shift from subitizing to counting. And subitizing and actual counting are associated with a different cognitive processing, subjective experience, and brain activation.

Recently, psychologists have discovered a link between the accuracy of people's ability to come up with approximate numbers and their overall mathematical ability. Justin Halberda of Johns Hopkins University tested fourteen-year-old children for their ability to estimate the number of rapidly flashing blue and yellow dots on a video screen. Since the dots, mixed like confetti on the screen, appeared too fast for deliberate counting, the children had to depend on what Halberda describes as "their approximate number system, their intuitive sense." The results were both surprising and thought provoking. After playing the video game for only ten minutes, a child's dot-counting prowess correlated with his performance on math tests from kindergarten up to the time of testing. Halberda believes, first, that it is possible to "train up" a person's number approximation ability; and second, that trained improvement will lead to a corresponding improvement in general mathematical ability. And as I've mentioned, mathematical proficiency is an important component of general intelligence.

If you aren't that proficient in math and, until now, didn't know why you should work at improving your proficiency, here are several kinds of puzzles that can help you do that.

DOT COUNT *(Mathematics)*

ANSWERS ON PAGE 250

▶ This puzzle strengthens your understanding of numbers and number patterns.

How many dots in each of these pictures? To count the dots efficiently, you must look for shortcuts.

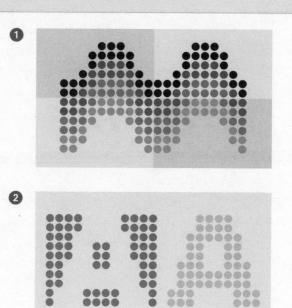

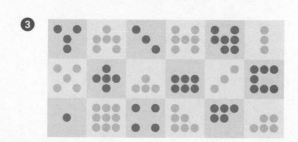

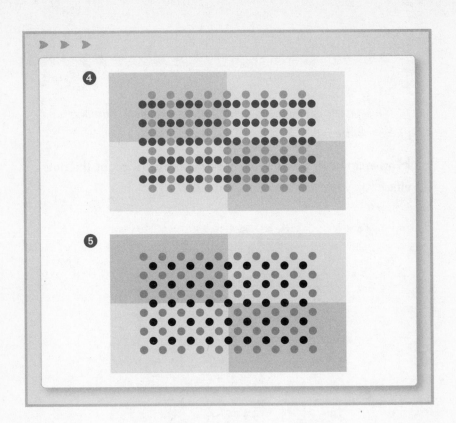

DOT SEQUENCE *(Mathematics)*

ANSWERS ON PAGES 251–254

▶ *This puzzle strengthens your working memory.*

In each puzzle below, there are three dot patterns that form a logical sequence. Draw the dot pattern that comes next in the sequence, then count how many dots it contains. For instance, in the first sequence, each pattern contains 4 more dots than the previous pattern, so the next pattern will contain 16 dots.

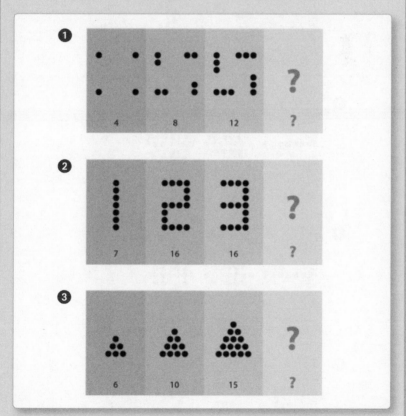

L

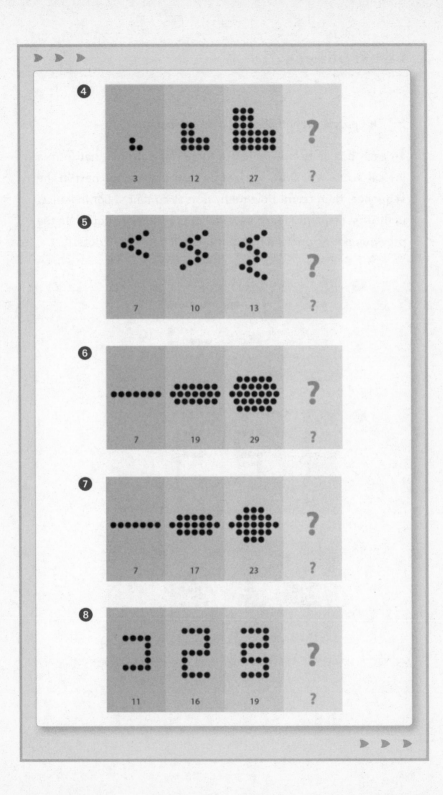

4

3   12   27   ?
?

5

7   10   13   ?
?

6

7   19   29   ?
?

7

7   17   23   ?
?

8

11   16   19   ?
?

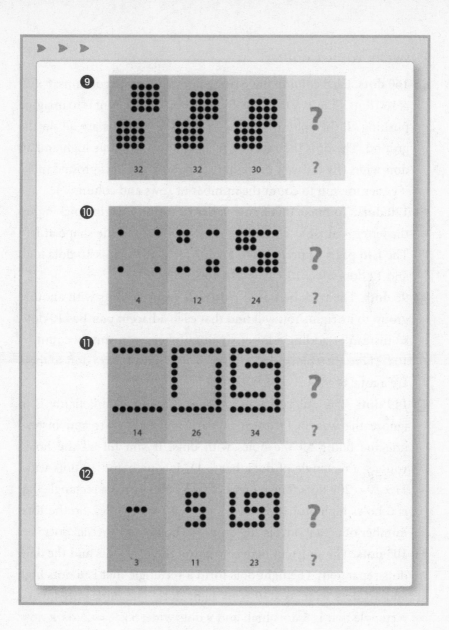

9   32   32   30   ?   ?

10   4   12   24   ?

11   14   26   34   ?

12   3   11   23   ?

ANSWERS TO DOT COUNT

❶ 160 dots. Each column has 8 dots, and there are 20 columns: 8 x 20 = 160 dots. One way to make this figure easier to count is to imagine pushing all the columns down so that the bottoms are all on the ground. The dots then form a rectangle that is 8 dots high and 20 dots wide. (By the way, the light rectangles in the background make it easier for you to count the number of rows and columns.)

❷ 120 dots. To make this figure easier to count, imagine picking up the letter A at right and fitting it into the hole in the shape at left. The two parts fit perfectly to form a rectangle that is 10 dots high and 12 dots wide: 10 x 12 = 120 dots.

❸ 93 dots. The trick here is to pair each group of dots with another group to its right. You will find that each adjacent pair has 10 dots, so instead of adding a lot of small numbers, you simply count by 10s. There are 9 pairs, and 3 of the groups have 1 extra dot, so there are a total of 90 + 3 = 93 dots.

❹ 149 dots. The quickest way to count the dots in this figure is to ignore the weaving pattern of dark and light dots and instead imagine filling all the holes with dots. If you fill all the holes, you get a rectangle of dots that is 11 dots high and 19 dots wide: 11 x 19 = 209 dots. The pattern of holes also forms a rectangle that is 6 holes high and 10 holes wide: 6 x 10 = 60 holes. So the total number of actual dots is 209 dots – 60 holes = 149 actual dots.

❺ 105 dots. The shortcut here is to count the light dots and the dark dots separately. The light dots form a rectangle that is 6 dots high and 10 dots wide: 6 x 10 = 60 light dots. The dark dots form a rectangle that is 5 dots high and 9 dots wide: 5 x 9 = 45 dark dots. So the total is 60 light dots + 45 dark dots = 105 dots.

REFLECTION. Which of these puzzles were easy to solve? Which were hard? How did you solve them?

ANSWERS TO DOT SEQUENCE

1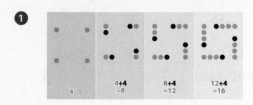

Each of the four lines gets one dot longer at each step.

2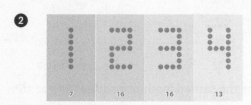

Each figure shows a number made out of dots. Each number is 7 dots high and 4 dots wide, except for the number 1.

3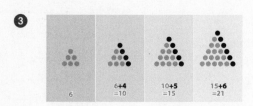

Each step adds a row of dots on the right edge. The number of dots added grows by 1 dot at each step.

4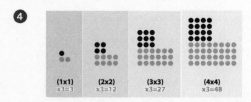

Each figure is made of three squares of the same size. The side length of the square grows by 1 dot at each step.

5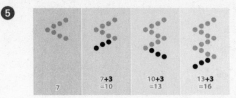

At each step, a new line of 3 dots is added.

6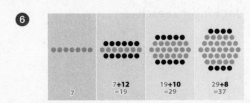

At each step, two new lines of the same length are added, one at the top and one at the bottom. The length of the lines decreases by 1 dot at each step.

7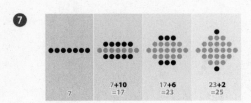

At each step, two new lines of the same length are added, one at the top and one at the bottom. The length of the lines decreases by 2 dots at each step.

8

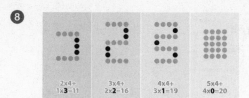

Each figure is made of several rows of 4 dots each, connected by vertical lines of dots. The number of rows increases by 1 dot at each step. The length of the vertical lines decreases by 1 dot at each step. In the last figure, the length of the vertical lines is zero, so there are no vertical lines separating the rows.

9

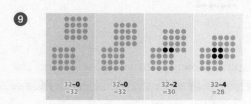

There are two 4-by-4 squares of dots. At each step the top square moves down 1 dot. The total number of dots is 16 + 16 minus the number of dots where the two squares overlap. In the last figure, the two squares overlap by 4 dots, so the total number of dots is 16 + 16 − 4 = 28.

10

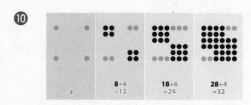

There are two patterns here. Two of the dots grow into horizontal lines, which get 1 dot longer at each step. The other 2 dots grow into squares, which get 1 dot wider at each step. In the last figure, the lines and squares overlap.

⑪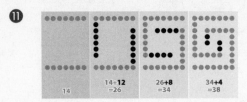

The pattern here is two lines spiraling in toward the center. The number of dots added to each spiral at each step keeps decreasing by 2.

⑫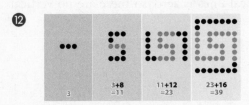

The pattern here is a double spiral growing out from the center. The number of dots added to each arm of the spiral at each step keeps increasing by 2.

REFLECTION. Which sequences were hard to figure out? Why? Which sequences helped you figure out other sequences?

ILLUSIONS: SHADOWS, BALLS, AND ROTATING SNAKES

LOOK AT the two shapes below, at left.

Now look at the framed image.

Which shape is present in the framed image, the top shape or the bottom one?

While most people have no difficulty spotting the top shape, fewer notice that the bottom shape is also contained in the framed image as part of the background. That failure in recognition occurs because the brain is specialized for perceiving figures rather than background.

Thus, the top shape as the foreground figure is readily detected, but the bottom shape, as part of the background, isn't so readily recognized. It's as if that background shape doesn't perceptually exist. "What is perceived as figure and what is perceived as ground do not have shape in the same way. In a certain sense, the ground has no shape," as the Danish physiologist Edgar Rubin described the relationship.

If a reversal of figure and background occurs, the brain's response changes accordingly. In an experiment demonstrating this, monkeys were trained to respond to a foreground shape, signaled by the firing of brain cells in the monkey's inferior temporal cortex, known to be involved early in shape perception and object recognition. But if a foreground/background reversal occurred (the background shape becomes the figure and the figure shape sinks into the background), the cells stopped firing and the monkey failed to respond to the shape. Similar figure/ground segregation occurs in the human brain.

Fix your gaze on the well-known vase/profile illusion created in 1915 by Edgar Rubin. You may see this ambiguous figure/ground illusion as either two black profiles looking at each other in front of a white background or as a white vase outlined against a black background. Some people have to stare at the illusion for several minutes before their brain performs the figure/background switch. Some even require descriptive instructions ("Picture the lip of the cup corresponding to the foreheads of the two profiles") before they're able to see both images. But no matter how quickly they shift from one image to the other, they can't see both of them at the same time.

Try it for yourself. You can't simultaneously see vase and profiles, because your brain encodes objects primarily in terms of contours, which, in the vase/profile illusion, change according to which of the two interpretations of the illusion you perceive at a given moment. In this illusion, the wavy lines extending down either side of the image can be seen as either human profiles (foreheads, noses, mouths, chins, and necks) or as a vase (lip, bowl, stem, and base). What you see depends on which side of these lines your brain focuses on to form the figure. Concentrate laterally (outwardly) and you see the faces; concentrate medially (inwardly) and you see the vase. With each switch in perspective, figure and background shift between the two possible interpretations. Correspondingly, an array of electrodes within your inferior temporal (IT) cortex would show different firing patterns as one interpretation of the illusion replaces another.

Another important point: The illusion isn't created just by the firing of arrays of cells within the inferior temporal cortex but must involve higher cortical processing. In order to recognize a vase as one possible interpretation of the illusion, you must have some experience with vases. Thus, this simple illusion involves brain processing extending from the most fundamental level (specific neurons involved in figure/ground perception) all the way to the association cortex, which stores knowledge of cultural artifacts. This knowledge will differ from one person to another. Some will see a vase suggestive of an artifact from the Edo period in Japanese history, while others will see just a vase.

In case you're still not convinced that you can't simultaneously perceive two interpretations of an illusory figure, stare for a minute at the two ambiguous depth figures shown on page 258. The first, known as the Necker cube, was designed in 1832 by Swiss crystallographer Louis Necker. In this well-known illusion, the two-dimensional lines of the figure are interpreted as the projection of a three-dimensional object, the cube. Because of an absence of depth clues, the figure remains ambiguous, i.e., capable of being seen from two different perspectives. In one interpretation, panel A is to the front and shifted slightly to the left. In the second interpretation, part of A forms the bottom of the cube as seen from below. Keep staring at the illusion and you will experience your brain shifting between the two perceptual interpretations. And while it's possible after some practice to alternate rapidly back and forth between the two interpretations, you cannot perceive them simultaneously: the brain is forced to interpret the ambiguous drawing only one way at a time.

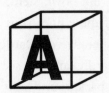

When you stare at the Necker cube, neither of the two interpretations involves a sense of movement on the page: the image remains stable, with the positional changes seeming to occur at distinct moments within the brain of the person staring at the illusion. In contrast, other illusions, such as the optimized Fraser-Wilcox illusion by Akiyoshi Kitaoka, below (known as the Rotating Snakes illusion), create a sense of movement on the page even though the image remains static. Your eye movements or blinks produce the fleeting perception of motion based on timing differences in the brain's processing of contrast: since black and white are higher contrast than dark gray and light gray, they elicit faster responses in the brain's visual system. These timing-response differences in the firing of select cells within the brain create (i.e., are interpreted as) a sense of motion. The motion disappears if you stare at the illusion without blinking or moving your eyes.

Timing differences in the firing rates of various networks of neurons are likely responsible for the fluctuations we experience in our interpretation of many illusions. In the Necker cube illusion, the activation of one or more networks (responsible for one interpretation) is rapidly replaced by activation of other networks (the basis for the rapid shift to the alternative interpretation). In the Rotating Snakes illusion, our perception doesn't shift at all once the snakes start moving coincident with our eye blinks or eye movements. Only by inhibiting these blinks or eye movements can we shut off the perception of the snakes' movements on the page. You can inwardly distinguish these two processes for yourself by noting the differences in your subjective experiences as you interact with the two illusions.

One final point about visual illusions: small differences in background can lead to major changes in perception, especially when motion is involved. For an example, go to the wonderful website of Illusionworks, at www.illusionworks.com. As these illusions make clear, our interpretations can be strikingly altered by small background changes.

Equivalents of optical illusions exist in the cognitive sphere as well. Unless deliberate efforts are made to avoid falling prey to them, our brains naturally tend to process language, logic, and thought in ways that lead to cognitive illusions, defined by Australian meteorologist Neville Nicholls as "analogous to optical illusions in leading to errors we commit without knowing we are doing so, except that they arise from our difficulties in quantifying and dealing with probabilities, uncertainty, and risk." Further, these cognitive illusions often lead to serious deviations from rational thinking, usefully delineated by University of Pennsylvania psychologist Jonathan Baron as "whatever kind of thinking best helps people achieve their goals."

Cognitive illusions are the inevitable result of the limited capacity of the human brain to overcome intuitive, oversimplified approaches to complex problems. The illusions result because the inappropriate use of simple rules of thumb—more formally dubbed heuristics—leads to biases and costly errors, especially in ambiguous situations where probabilities are involved.

Consider the following two examples:

In the first example, you are given $300, along with the opportunity to increase your winnings either by receiving another $100 (totaling $400) or by tossing a coin. If you win the toss, you get another $200 (totaling $500); if you lose the toss, you get nothing (back to the original $300). What would you do?

In the second example, you are given $500 and a choice between immediately surrendering $100 (reducing your amount to $400) or tossing a coin. If you lose the toss, you have to return $200 (reducing your take to $300); if you win, you don't have to return anything. What would you do?

In the first example, most people elect to forgo the toss and pocket the $400, preferring a sure gain. In the second example, most people will choose to toss the coin, thus perhaps ending up with $100 less than they would have had by simply accepting the initial $100 loss. Why this difference?

For most of us, the threat of a loss exerts a greater effect on our decisions than the possibility of an equivalent gain, according to tests and surveys gauging this. For example, the distress we experience after losing $100 on the street far exceeds the pleasure that accompanies finding that same amount. Mirroring this experience, measurements of brain activity show what neuroscientists refer to as a "negativity bias" wherein negative information (either in the form of pictures or words) exerts a more powerful effect on the brain than positive information. But this bias can be overcome simply by reordering the sequence in which information is presented. If I describe someone you are about to meet as "intelligent, industrious, impulsive, critical, stubborn, envious," you're likely to form a more favorable opinion of him than you would if I reversed the order to "envious, stubborn, critical, impulsive, industrious, intelligent," even though the second list is simply the reverse of the first and presents the same information.

A similar ordering-of-information effect occurred in the two scenarios just described. In the first you are given the opportunity to "increase your winnings," whereas in the second the wording ("immediately surrendering") plays on your brain's tendency to avoid loss along with its negativity bias by requesting that you incur a loss by returning $100 from the $500 you were just given. To avoid that dreaded loss, most people—even those not especially given to risk taking or gambling—will toss the coin and perhaps lose even more.

Whenever a stockbroker advises a client to sell a stock that continues to fall in value, he must factor into his calculations the client's intolerance for losses. The client's hope that the stock will rebound and regain its value (resulting in a loss if sold before this happens) interferes with the sober realization that some stocks are never going to regain value and a limited loss now is preferable to a bigger loss later.

Think of the situations described above as two perceptions of a Necker cube in which one perception (fear of losses) ordinarily exerts a greater influence on our behavior than the other perception (satisfaction from gains). Just as you can increase your ability to shift back and forth between two perceptions of the Necker cube, you can increase your tolerance for loss simply by keeping in mind your brain's inherent tendency to misperceive and misinterpret situations in which loss is a potential outcome. As an alternative, you can hire a professional to do it for you.

Another cognitive illusion determines how we interpret information that conflicts with our preconceptions. In a classic study, two groups were tested on the certitude of their beliefs about capital punishment. The first favored capital punishment; the second believed the practice should be eliminated. Both groups were fiercely adamant about their opinions. All of the participants in the study read a report claiming to prove that capital punishment effectively deterred violent crime along with a second equally credible report claiming the practice didn't have any effect. From a strictly logical point of view, reading such balanced evidence should have led at least some people in each group to express a tiny decrease in their confidence in the correctness of their beliefs. Instead, the subjects expressed even greater confidence than they had before reading the balanced reports. How to explain such a finding?

"Strong initial views are resistant to changes because they influence the way that subsequent information is interpreted," according to Paul

R

Slovic, one of the world's experts on risk and risk perception, writing in his classic *Science* paper "Perception of Risk." "New evidence appears reliable and informative if it is consistent with one's initial beliefs; contrary evidence tends to be dismissed as unreliable, erroneous, or unrepresentative."

Illusions also occur in the exercise of memory. Write out on a piece of paper or type on your computer screen the following list of words: *bed, rest, tired, snore, dream, wake, nap, snooze, blanket, pillow, pajamas, robe, alarm clock, sheet,* and *mattress.* Read them aloud and then after five minutes write down as many of the words as you can remember. Finished? Now read over your list.

If you're like many people taking this test, you included the word *sleep* on your list even though it was not on the original list. Psychologists around the world have duplicated this finding that people frequently fall victim to the "associative memory illusion" whereby words associated with a "critical item" (in this instance *sleep*) lead to the false memory that the word appeared in the original word list.

Associative memory illusions result from the brain's tendency to make sense of seemingly random events—the "interpreter" effect of Michael Gazzaniga, as described earlier. Since all of the words on the above list can be related to sleep, the word *sleep* is assumed to have been on the list, even though it wasn't.

A related cognitive illusion is the "causal inference" illusion. For example, imagine yourself looking at several pictures including one of oranges lying haphazardly on a supermarket floor. Sometime later you're shown a picture of someone reaching out for an orange from the bottom of a stack and you're asked if you have seen that picture before. When asked this question, a statistically significant number of people experience the causal-inference illusion and respond that they have seen the picture before even though they haven't. While correctly inferring the most likely explanation for the clutter of oranges on the floor, their memory of having seen a picture of someone causing the tumbling of the oranges is an illusion created by the confusion between what was seen and its cause.

Here is a series of puzzles that will test your ability to detect and counter the influence of cognitive illusions:

COGNITIVE ILLUSIONS *(Focus, Mathematics)*

ANSWERS ON PAGES 265–266

▶ *This puzzle challenges you to think clearly in the face of distracting influences.*

COGNITIVE ILLUSIONS

❶ Heavy Subjects

Which weighs more, a pound of feathers or a pound of lead?

❷ Probable Causes of Death

Next we turn to cognitive illusions created by emotional reactions. We all know that the chances of dying in an airplane crash are much less than the chances of dying in a car crash over the same distance. But many people fear flying so much that they would rather drive than take a plane. Can you match each of the following possible causes of death with the odds of it occurring?

1. Bicycle accident	a.	1 in 5
2. Drowning	b.	1 in 84
3. Falling	c.	1 in 218
4. Fireworks	d.	1 in 1,008
5. Flood	e.	1 in 4,519
6. Heart disease	f.	1 in 13,729
7. Hot weather	g.	1 in 79,746
8. Lightning	h.	1 in 144,156
9. Motor vehicle accident	i.	1 in 340,733

▶ ▶ ▶

A

▶ ▶ ▶

❸ *The Monty Hall Problem*

Now get ready for the all-time champ of cognitive illusions. It's a probability stumper called the Monty Hall problem, named after the host of the television game show *Let's Make a Deal*. It describes a situation that Monty Hall actually presented to many contestants. I was certainly fooled the first time I heard this one. And even after you read the explanation of the correct answer, I guarantee that many of you will be absolutely convinced that I am wrong.

You're a contestant on Monty Hall's *Let's Make a Deal*. There are three doors onstage. Hidden behind one door is a new car; behind the other two are goats. You choose Door 1.

A. What are the odds that the car is behind your door?

B. Monty opens the other two doors to reveal goats. Now what are the odds that the car is behind your door?

C. Now the Monty Hall paradox: Monty opens Door 3, revealing a goat. He offers you a deal: Keep your door, or trade it in for the other unopened door. Should you trade? To guide you toward the right reasoning, also consider the following variation. There are a hundred doors. Behind one of the doors is a car; all the other doors conceal goats. You choose one of the doors, and Monty opens ninety-eight of the remaining

▶ ▶ ▶   ninety-nine doors to reveal goats. He offers you a deal: Keep your door, or trade it in for the other unopened door. Should you trade?

D. Here is a variation on the Monty Hall paradox: There are three doors. Behind two are cars; behind the third is a goat. You choose Door 1, and Monty opens Door 3 to reveal a car. He offers you a deal: Keep your door, or trade it in for the other unopened door. Should you trade?

ANSWERS TO COGNITIVE ILLUSIONS

❶ **HEAVY SUBJECTS**

After a moment of thought, you can probably answer this question correctly, but it is awfully tempting to answer that the pound of feathers weighs less. Of course it does not—both weigh a pound—but the feathers are less dense and softer, qualities we associate with lighter objects.

The size/weight illusion becomes far more compelling when you are presented with an actual bag of feathers and a bar of lead. Scientific experiments show that as many as 98 percent of subjects misjudge the weight relationship when the lighter object is smaller and denser.

❷ **PROBABLE CAUSES OF DEATH**

Heart disease	1 in 5
Motor vehicle accident	1 in 84
Falling	1 in 218
Drowning	1 in 1,008
Bicycle accident	1 in 4,519
Hot weather	1 in 13,729
Lightning	1 in 79,746
Flood	1 in 144,156
Fireworks	1 in 340,733

❸ MONTY HALL PROBLEM

A. There is a one-in-three (33 percent) chance that your door has the car behind it.

B. Since both other doors have goats, the chance that your door has the car is 100 percent.

C. Surprisingly, it is better to trade. When you chose Door 1, there was a one-in-three chance that the car would be behind it. This leads to two possibilities. Possibility 1: The car is behind Door 1, in which case the host can choose to open either of the other two doors. In this case, which occurs one-third of the time, it is better to stick with Door 1 and not trade, since the car is behind Door 1. Possibility 2: The car is not behind Door 1, in which case the host has no choice but to open the one closed door with the goat. In this case, which occurs two-thirds of the time, the host's choice of door tells you that the car is behind the door that he did not open, so it is better to trade doors. In other words, one-third of the time it is best to stay, and two-thirds of the time it is better to trade. The Monty Hall problem is deceptive because it seems that the host's choice of door doesn't tell you anything useful. Some of the time it doesn't, but two-thirds of the time it does.

D. You should *not* trade. In this variation on the Monty Hall paradox, the roles of cars and goats have been reversed. By opening a door to reveal a car, the host increases the odds that the remaining closed door hides a goat to two in three. Assuming you want a car, not a goat, you shouldn't trade.

REFLECTION. Which of these puzzles were you able to solve? What misled you? Do you believe my answers?

14

CREATIVITY: THE MAGIC MATCHES OF CARLO REVERBERI

WHILE THE FRONTAL LOBES are a great asset when it comes to our overall mental functioning, there are times when they can impose restraints on our logical and problem-solving abilities. To experience this, here is a series of problems created by the brilliant psychologist Carlo Reverberi, which involve the simple rearrangement of matchsticks.

MATCHSTICKS *(Creative Problem-Solving)*

ANSWERS ON PAGE 269

▶ *On the next page, there is a fiendishly clever mental exercise. Arrange matchsticks according to a given incorrect equation and then move any one of the matchsticks to come up with a correct equation. Three rules govern your choices: Only one stick can be moved; no stick can be discarded; and the final arrangement must result in a correct mathematical statement. Move as quickly as you can from one to the other, and time how long it takes you to solve each problem.*

▶ ▶ ▶

1. $IV = III - I$

2. $VI = VII + I$

3. $III = III + III$

4. $IV = III + III$

5. $V = III - II$

6. $VI = VI + VI$

7. $VIII = VI - II$

8. $IV = IV + IV$

9. $II = III + I$

10. $VII = VII + VII$

11. $VII = II + III$

12. $VI = IV - II$

1. IV = III = I
2. VII = VII + I
3. III = III ≠ III
4. VI = III + III
5. V = III = II
6. VI = VI ≠ VI
7. VIII = VI = II
8. IV = IV ≠ IV
9. III = II + I
10. VII = VII ≠ VII
11. VI = II + III
12. VI = IV = II

Most people find the third, sixth, eighth, and tenth the most difficult. In order to understand why, consider that the matchstick problems constitute three different classes:

In the first, you move a matchstick that forms part of one numeral in order to make another numeral, i.e., VII = II + III is solved by moving a matchstick from VII to II, yielding VI = III + III. Most people find these exercises easy.

In the second class of problem, you move a matchstick from an equal sign to a minus sign, thus converting it into an equal sign, i.e., V = III−II converts to V−III = II. These are of moderate difficulty.

In the final class of problem, you change a plus sign by rotating its vertical component through 90 degrees to create an equal sign, i.e.,

VIII = VII + VII converts to VII = VII = VII. This third type most likely took you the longest to successfully solve. Why? Because such problems require the puzzle solver to think beyond the typical form of an equation in which the variables are of two types. In the first, numbers alone are involved (Class I problems). Second, the change requires converting an equal sign to a minus sign by mentally "picking up" the matchstick and moving it from one part of the equation to another (Class II problems).

The change required for successfully solving the Class III problem is subtler and involves a 90-degree rotation of only one element of the equation.

While most people are better at solving Class I and Class II problems than Class III problems, the exact opposite holds for patients with injuries or disease-related damage to their frontal lobes. They lag behind individuals with normal frontal lobes on Class I and Class II problems but consistently outperform them on Class III problems. Only 43 percent of healthy controls in Reverberi's puzzle experiment could solve the most difficult Class III matchstick problems, compared with 82 percent of people who had damage to their lateral frontal lobes. What is the explanation?

One explanation is that the lateral frontal lobes bias our responses toward those aspects of a problem or a puzzle that we're most familiar with. Healthy frontal lobes focus increased attention to numbers rather than to operations (+ and −) or to alterations of an equal sign (=). As a result, people with normally functioning frontal lobes are much less comfortable working a statement containing two equal signs, which is a situation not commonly encountered when working algebraic equations.

Reverberi's matchstick puzzles point out the need to defend ourselves against what psychologists refer to as "functional fixedness": a mental block against using an object in a new way that is required to solve a problem. If we don't take proper precautions, our frontal lobes can on occasion lock us into perspectives and interpretations that are less than optimal. This isn't meant to downplay the contribution of the frontal lobes: on the contrary, the frontal lobes are the engine that has enabled our species to move ahead of every other species on the planet. But sometimes we adopt too literal an approach when we actually should be doing what Reverberi refers to as addressing a problem with "a trial-and-error approach without prior assessment of the likely fruitfulness or appropriateness of a strategy." In other words, don't be afraid to "think stupid": assume nothing in order not to overlook the most seemingly "obvious" approach to a problem.

Sometimes puzzles are solved not by the executive processing best carried out by the frontal lobes (planning, simulating, deciding, evaluating, etc.) but by means of a sudden insight: one moment you have no idea how to solve a puzzle; the next moment, the correct answer comes to you "in a flash." Several brain regions become active during insight. The anterior temporal lobes are especially important.

When listening to a joke, the anterior temporal areas become active at the moment we correlate the introductory patter with the final punch line. It is also active when we read a story or try to come up with a verbal description of a unique occurrence or series of circumstances. So it shouldn't be surprising to learn that sudden insights are also marked by anterior temporal activation. The time sequence is especially intriguing. About a third of a second prior to the realization of sudden insight, increased brain wave activity can be detected within the anterior temporal area. In other words, the experience of sudden insight is preceded by anterior temporal brain activation prior to the conscious recognition of the solution to the puzzle. So who should be given credit for the solution to the puzzle?

Consider this riddle: What would be the worst day of the year to be told that you had just won a large amount of money?

While you first read and ponder such a riddle, an fMRI would show activity in a broad swath of brain tissue, including the lateral prefrontal, the posterior parietal, the cingulate, and the anterior temporal areas. In many cases, this network will provide you with the answer. But when the puzzle eludes solution by logical inference (a kind of silent reasoning to oneself), those first three areas become less important. However, such an approach is unhelpful in this riddle. (The answer has nothing to do with taxes, as most people immediately and mistakenly assume.) The solution emerges by means of the sudden insight provided by the anterior temporal areas. Since this is an entirely different way to solve a puzzle, it's accompanied by a unique subjective sensation: the joy of sudden and, until that moment, unanticipated insight. In this riddle every day of the year would be great to be told you had just won money except for one date: April 1—April Fool's Day! (If you're from a country where April Fool's Day isn't part of the culture, you couldn't be expected to come up with the answer from personal experience.)

Here are several other puzzles that largely resist overintellectualized approaches but are frequently solved by sudden insight. At the moment you suddenly gain insight into the puzzles, you'll subjectively experience your temporal lobe "kicking in."

MORE MATCHSTICKS
(Spatial Thinking, Mathematics, Creativity)

ANSWERS ON PAGE 274

▶ *These puzzles challenge you to think creatively about spatial and mathematical problems. However, they require different insights from those used in the previous matchstick puzzles.*

❶ Use 15 matchsticks to make the figure above. Remove 6 matchsticks and leave 10.

▷ ▷ ▷

② Use 10 matchsticks to spell the word FIVE as shown above. Remove 7 matchsticks and leave 4.

③ Use 10 matchsticks to spell the word TWO as shown above. Move 5 matchsticks to make 21.

④ Use 12 matchsticks to make the incorrect equation shown above. Without moving any matchsticks, how can you make this equation correct?

ANSWERS TO MORE MATCHSTICKS

1 Remove 6 matchsticks to leave the written-out word TEN.

2 Remove the 7 matchsticks in the letters F and E to leave the Roman numeral for 4.

3 Move the 5 matchsticks in the T and most of the O to make the Roman numeral for 21.

4 Walk around the table and look at the equation upside down from the other side.

So far, we have looked at creativity in the context of problem solving, where a flash of insight leads to a surprising predetermined solution. Here are puzzles that explore more open-ended forms of creativity in which there is no one right solution.

DIVERGENT THINKING *(Visual Thinking, Creativity)*

▶ *This activity challenges you to think flexibly and creatively.*

These classic creative-thinking exercises challenge you to re-interpret a familiar object many different ways. As with the exercise where you are challenged to generate lots of words starting with the same letter, the challenge here is to let your mind wander freely through many possibilities without getting stuck on any one.

10 USES FOR A BRICK

As quickly as you can, think of 10 different uses for a brick. Try to make the uses as different as possible. For instance, don't just list 10 things you can build by stacking bricks. Once you have 10 uses, think of 10 more.

▶ ▶ ▶

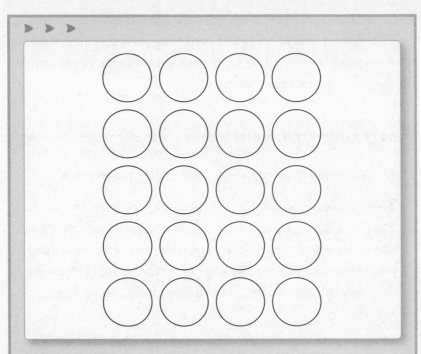

20 CIRCLES

Draw lines to turn these 20 circles into pictures that are as different as possible from one another. Don't simply draw 20 different coins.

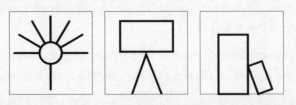

DROODLES

Droodles are ambiguous abstract cartoons with humorous captions. What is the first droodle above, on the left? A frightened mop? A spider doing a handstand? Invent three other humorous captions for this droodle. Then invent at least three humorous captions for each of the other two droodles.

ANSWERS ON PAGE 282

▶ *This activity challenges you to think flexibly and creatively.*

One of my favorite creative activities is making ambigrams. An ambigram is a word written so it can be read in more than one way. For instance, here is an ambigram I created on the title of this book. Notice that the design is perfectly symmetrical: it reads the same both right side up and upside down.

the playful brain

A few words in English, like NOON and SWIMS, are natural ambigrams.

NOON SWIMS

In the 1970s, artist John Langdon and I independently hit on the idea of modifying words that are not naturally symmetrical so they have perfect symmetry. Both of us found that creating ambigrams is a challenging and rewarding pursuit. Cognitive scientist and ambigram artist Douglas Hofstadter coined the word *ambigram*. Ambigrams now enjoy worldwide popularity, fueled by their appearance in Dan Brown's novel *Angels & Demons*.

▶ ▶ ▶

A

▶ ▶ ▶

Here are some creative exercises that will invite you to make ambigrams of your own.

CHUMP

Get paper and a pencil. Write the word CHUMP in lowercase cursive letters, with all of the letters connected. Don't quite close the bowl of the final P: Leave it slightly open. Be sure all of the letters are lowercase. When you are done, turn the paper upside down. Congratulations! You have just created your first ambigram.

USA

The word CHUMP happens to work easily in cursive. Here is a word that takes a bit of work to turn it into an ambigram. Figure out a way to write the word USA so it reads the same right side up and upside down. You will notice that the central S is naturally symmetrical, but you will have to modify the other letters so the initial U turns into an A. Keep turning your design around to make sure it reads equally well in both directions. When you have found a solution, try finding a different solution. There are quite a few different ways to make this ambigram work.

VISTA

Here is a slightly harder ambigram. Write the word VISTA so it reads the same right side up and upside down. Again, the central S is naturally symmetrical, but this time there are two pairs of letters to modify: V into A, and I into T. There are many solutions to this ambigram. Try finding at least two.

CHILD

Find two different ways to turn CHILD so it reads the same right side up and upside down. The straightforward way is to turn the C into D, the H into L, and the I into I. But you can also turn one letter into two, such as H into IL. This technique is called letter regrouping.

YOUR NAME

Find a way to write your first name so that it reads the same right side up and upside down. When you are done, show your creation to other people and ask if they can read it. Of course, some names work more easily than others. If your name is OTTO, you are in luck; if your name is CHRISTOPHER, you have a challenge on your hands. If you get stuck, feel free to choose a nickname, last name, or other form of your name—or the name of a friend.

EXAMPLES OF AMBIGRAMS

I could give more detailed instruction on how to create ambigrams—and I will in a moment—but half the fun is discovering techniques on your own. See how far you can get with your own name before reading further.

Here are a few more ambigrams to inspire you. Notice the use of a mixture of uppercase and lowercase letters, and that one letter sometimes becomes more than one letter when turned upside down.

heather

krick

(Garth)

Dolores

Ashley

tony

▷ ▷ ▷

TIPS FOR MAKING GOOD AMBIGRAMS

Write it twice. Start by writing your name in ordinary letters. Then turn the page upside down and write it again, just below the first name, so the two copies of the name align with the first letter of one copy just below the last letter of the other copy.

Look for compromises. Next, look for ways to modify the letter shapes in one copy of the name so they look more like the upside-down letter in the other copy of the name. Some letters will fall in place easily, like S into S or p into d, while others will be more of a struggle. Keep turning your design around to make sure it reads equally well in both directions.

Try many letter forms. Don't get stuck on one form of a letter. If uppercase doesn't work, try lowercase. Also, try cursive and ornate forms of letters. Remember that lowercase *a* has two different forms: the one-story **a** we all learned to print in school, and the two-story **a** used in books.

Regroup letters. Instead of always turning one letter into one letter, consider turning one letter into two letters. For instance, turning N into I is hard, but turning N into IV is easy.

Polish the style. Once you have worked out the basic lines of an ambigram, try dressing it up with a font style. Borrow ideas from fonts in magazines, books, and computers. Try thickening the strokes, adding serifs (the little spurs at the ends of strokes), and other embellishments.

Test for legibility. Strive to make your ambigram as easy to read as possible. Legibility is an important and elusive quality for ambigrams. It is easy to fool yourself into believing that an illegible ambigram is easy to read, because you know what it says. So show your ambigrams to other people.

THE PLAYFUL BRAIN

CHUMP. Here is the ambigram on CHUMP.

chump

According to *Ripley's Believe It or Not!* CHUMP is the only word in English that works naturally as an ambigram when written in cursive letters. That isn't quite right: the P has to be altered slightly to make the ambigram work, and the unusual word MU (it means a letter of the Greek alphabet) works perfectly without modification. But in any case, natural cursive ambigrams are rare.

USA. Here are several different solutions. Notice that you can either add the crossbar to the U or remove it from the A. Also, notice the different ways the spurious crossbar is rationalized visually.

USA USA USA
USA USA USA

VISTA. Here are two different solutions, using uppercase and lowercase letters.

VISTA *vista*

CHILD. Here are three solutions. Notice how the letters are grouped differently in different versions: H turns into L in the first ambigram, while H turns into IL in the second ambigram, and CH turns into LD in the third ambigram. Which of these ambigrams do you find easiest to read?

CHILD CHILD CHILD

CONCLUSIONS

IF SCOTT AND I have succeeded, you have learned a lot about the brain and have done so in an instructive, entertaining, and, we hope, even delightful manner: we've challenged you to challenge yourself with puzzles that exercise different parts of your brain. Now you can look at puzzles in new ways: not as struggles to be slogged through but as opportunities to expand and exercise your brain.

Rather than just another book describing the brain, we wanted your first brain-puzzle adventure to be like opening a small confectioner's box and selecting a fine chocolate or other candy delight.

Once you have sampled the delights of *The Playful Brain*, you may hunger for more "brain candy." If so, head to your local library or bookstore and stock up on puzzle books. Turn to the puzzle section of the newspaper and try a puzzle you've never tried before. Scour your local toy store for puzzle toys. Pull out your board games and invite friends over to play.

You can also look online for brain-stimulating games to play on your computer or phone. There are many websites that provide brain

B

exercise programs. You'll find other suggestions for online brain games on our website, www.theplayfulbrain.com.

Throughout the writing of this book, we have kept in mind that neuroscientists are still in the very early stages of understanding the brain. Even though they have discovered more about the brain during the past three decades than in the previous three hundred years, wonderful and exciting discoveries lie ahead. As a response to these discoveries, Scott and I hope to work together on further explorations into the ways that you can use puzzles to help develop a more powerful brain. In the meantime, we welcome your comments and suggestions. E-mail me via my website, www.richardrestak.com, or Scott at scott@scottkim .com; you may also visit the book's website at www.theplayfulbrain.com.

ACKNOWLEDGMENTS

With grateful appreciation for the generous help and encouragement provided by my editor, Jake Morrissey; his editorial assistant, Sarah Bowlin; my agent, Sterling Lord; my brother Christopher Restak; and Barbara Best at the Dana Foundation.

R.R.

RESOURCES

Aarts, Henk, Ruud Custers, and Hans Marien. "Preparing and Motivating Behavior Outside of Awareness," *Science*, 319 (March 21, 2008), 1639.

Bays, Paul M., and Masud Husain. "Dynamic Shifts of Limited Working Memory Resources in Human Vision," *Science*, 321 (August 8, 2008), 851–854.

Bielock, Simon L., Thomas H. Carr, Claire MacMahon, and Janet L. Starkes. "When Paying Attention Becomes Counterproductive: Impact of Divided Versus Skill-Focused Attention on Novice and Experienced Performance of Sensorimotor Skills," *Journal of Experimental Psychology: Applied*, 8, no. 1 (March 2002), 6–16.

Cohen, M. S., S. M. Kosslyn, et al. "Changes in Cortical Activity During Mental Rotation: A Mapping Study Using Functional Magnetic Resonance Imaging," *Brain*, 119 (1996), 89–100.

Conway, Bevil R., et al. "A Neural Basis for a Powerful Static Motion Illusion," *The Journal of Neuroscience*, 25, no. 23 (June 8, 2005), 5651–5656.

Douglas, Kate. "The Subconscious Mind: Your Unsung Hero." *New Scientist*, December 29, 2007.

Eagleman, Douglas M., et al. "Time and the Brain: How Subjective Time Relates to Neural Time," *Journal of Neuroscience*, 25, no. 45, 10369–10371.

Epstein, Russell A. "Parahippocampal and Retrosplenial Contributions to Human Spatial Navigation," *Trends in Cognitive Sciences*, 12, no. 10, 388–396.

Fusi, Stefano. "A Quiescent Working Memory," *Science*, 319 (March 14, 2008), 1495–1496.

Galdi, Silvia, Luciano Arcuri, and Bertram Gawronski. "Automatic Mental Associations Predict Future Choices of Undecided Decision-Makers," *Science*, 321 (August 22, 2008), 1100–1102.

Karpicke, Jeffrey D., and Henry L. Roediger III. "The Critical Importance of Retrieval for Learning," *Science*, 319 (February 15, 2008), 966–968.

Kim, Scott. *Inversions: A Catalog of Calligraphic Cartwheels.* Key Curriculum Press, 1996.

Kim, Scott. Mind Benders and Brainteasers Puzzles Page-A-Day Calendar. Workman, 2010.

Lawton, Graham. "Mind Tricks: Six Ways to Explore Your Brain," *New Scientist*, September 19, 2007.

Lawton, Graham. "A Game to Train Your Brain? A Spell in the Brain Gym Is Supposed to Make Us Smarter," *New Scientist*, January 12–18, 2008, 26–29.

Macknik, Stephen, L., et al. "Attention and Awareness in Stage Magic: Turning Tricks into Research," *Nature Reviews/Neuroscience*, 9 (November 2008), 871–879.

Martines-Conde, Susana, and Stephen L. Macknik. "Magic in the Brain: How Magicians 'Trick' the Mind," *Scientific American*, November 24, 2008.

Mongillo, Gianluigi, Omry Barack, and Misha Tsodyks. "Synaptic Theory of Working Memory," *Science*, 319 (March 14, 2008), 1543–1546.

Motluk, Alison. "How Many Things Can You Do at Once?" *New Scientist*, April 7, 2007.

Parris, B. A., G. Kuhn, and T. L. Hodgson. "Imaging the Impossible: A Neuroimaging Study of Cause and Effect Violations in Magic Tricks," Society of Neuroscience, program number 262–21, October 15, 2006.

Phillips, Helen, et al. "Creative Minds," *New Scientist*, supplement, October 29, 2005, 39–54.

Polmann, Stefan, and Mariane Maertens. "Shift of Activity from Attention to Motor-Related Brain Areas During Visual Learning," *Nature Neuroscience*, 8, no. 11 (November 2005), 1494–1496.

Restak, Richard, M.D. *Mozart's Brain and the Fighter Pilot: Unleashing Your Brain's Potential.* Harmony Books, 2001.

Restak, Richard, M.D. *Think Smart: A Neuroscientist's Prescription for Improving Your Brain's Performance.* Riverhead Books, 2009.

Reverberi, Carlo, et al. "Better Without 'Lateral' Frontal Cortex? Insight Problems Solved by Frontal Patients," *Brain*, 128 (2005), 2882–2890.

Rubin, Nava. "Figure and Ground in the Brain," *Nature Neuroscience*, 4, no. 9 (September 2001), 857–858.

Schaffer, Karl, Erik Stern, and Scott Kim. *Math Dance with Dr. Schaffer and Mr. Stern.* Movespeakspin (mathdance.org), 2000.

Sheth, B., J. Bhattacharya, and D. Wu. "On the Neural Track of Eureka," Society for Neuroscience, program 138. 4, October 24, 2004.

Swaminathan, Nikhil. "What Are We Thinking When We (Try to) Solve Problems?" *Scientific American*, January 15, 2008.

Vinod, Goel, et al. "Hemispheric Specialization in Human Prefrontal Cortex for Resolving Certain and Uncertain Inferences," *Cerebral Cortex*, 17 (October 2007), 2245–2250.

INDEX

N